Graduate Texts in Mathematics 116

John L. Kelley T. P. Srinivasan

Measure and Integral

Volume 1

Springer-Verlag
New York Berlin Heidelberg
London Paris Tokyo

John L. Kelley
Department of Mathematics
University of California
Berkeley, CA 94720
U.S.A.

T.P. Srinivasan
Department of Mathematics
The University of Kansas
Lawrence, KN 66045
U.S.A.

AMS Classification: 28-01

Library of Congress Cataloging-in-Publication Data
Kelley, John L.
Measure and integral/John L. Kelley, T.P. Srinivasan.
p. cm.—(Graduate texts in mathematics; 116)
Bibliography: p.
Includes index.
ISBN 0-387-96633-1
1. Measure theory. 2. Integrals, Generalized. I. Srinivasan, T.P.
II. Title. III. Series.
QA312.K44 1988
515.4'2—dc19 87-26571

Typeset by Asco Trade Typesetting Ltd., Hong Kong.
Printed and bound by R.R. Donnelley & Sons, Harrisonburg, Virginia.
Printed in the United States of America.

9 8 7 6 5 4 3 2 1

ISBN 0-387-96633-1 Springer-Verlag New York Berlin Heidelberg
ISBN 3-540-96633-1 Springer-Verlag Berlin Heidelberg New York

PREFACE

This is a systematic exposition of the basic part of the theory of measure and integration. The book is intended to be a usable text for students with no previous knowledge of measure theory or Lebesgue integration, but it is also intended to include the results most commonly used in functional analysis. Our two intentions are some what conflicting, and we have attempted a resolution as follows.

The main body of the text requires only a first course in analysis as background. It is a study of abstract measures and integrals, and comprises a reasonably complete account of Borel measures and integration for $\mathbb{R}$. Each chapter is generally followed by one or more supplements. These, comprising over a third of the book, require somewhat more mathematical background and maturity than the body of the text (in particular, some knowledge of general topology is assumed) and the presentation is a little more brisk and informal. The material presented includes the theory of Borel measures and integration for $\mathbb{R}^n$, the general theory of integration for locally compact Hausdorff spaces, and the first dozen results about invariant measures for groups.

Most of the results expounded here are conventional in general character, if not in detail, but the methods are less so. The following brief overview may clarify this assertion.

The first chapter prepares for the study of Borel measures for $\mathbb{R}$. This class of measures is important and interesting in its own right and it furnishes nice illustrations for the general theory as it develops. We begin with a brief analysis of length functions, which are functions on the class $\mathcal{J}$ of closed intervals that satisfy three axioms which are eventually shown to ensure that they extend to measures. It is shown in chapter 1 that every length function has a unique extension λ to the lattice $\mathscr{L}$ of sets generated by $\mathscr{J}$ so that λ is *exact*, in the sense that $\lambda(A) = \lambda(B) + sup\{\lambda(C): C \in \mathscr{L} \text{ and } C \subset A \setminus B\}$ for members A and B of $\mathscr{L}$ with $A \subset B$.

The second chapter details the construction of a pre-integral from a pre-measure. A real valued function μ on a family $\mathscr{A}$ of sets that is closed under finite intersection is a *pre-measure* iff it has a countably additive non-negative extension to the ring of sets generated by $\mathscr{A}$ (e.g., an exact function μ that is continuous at $\varnothing$). Each length function is a pre-measure. If μ is an exact function on $\mathscr{A}$, the map $\chi_A \mapsto \mu(A)$ for A in $\mathscr{A}$ has a linear extension I to the vector space L spanned by the characteristic functions χ_A, and the space L is a vector lattice with truncation: $I \wedge f \in L$ if $f \in L$. If μ is a pre-measure, then the positive linear functional I has the property: if $\{f_n\}_n$ is a decreasing sequence in L that converges pointwise to zero, then $lim_n I(f_n) = 0$. Such a functional I is a *pre-integral*. An *integral* is a pre-integral with the Beppo Levi property: if $\{f_n\}_n$ is an increasing sequence in L converging pointwise to a function f and $sup_n I(f_n) < \infty$, then $f \in L$ and $lim_n I(f_n) = I(f)$.

In chapter 3 we construct the Daniell–Stone extension L^1 of a pre-integral I on L by a simple process which makes clear that the extension is a completion under the L^1 norm $\|f\|_1 = I(|f|)$. Briefly: a set E is called *null* iff there is a sequence $\{f_n\}_n$ in L with $\sum_n \|f_n\|_1 < \infty$ such that $\sum_n |f_n(x)| = \infty$ for all x in E, and a function g belongs to L^1 iff g is the pointwise limit, except for the points in some null set, of a sequence $\{g_n\}_n$ in L such that $\sum_n \|g_{n+1} - g_n\|_1 < \infty$ (such sequences are called *swiftly convergent*). Then L^1 is a norm completion of L and the natural extension of I to L^1 is an integral. The methods of the chapter, also imply for an arbitrary integral, that the domain is norm complete and the monotone convergence and the dominated convergence theorems hold. These results require no measure theory; they bring out vividly the fundamental character of M. H. Stone's axioms for an integral.

A *measure* is a real (finite) valued non-negative countably additive function on a δ-ring (a ring closed under countable intersection). If J is an arbitrary integral on M, then the family $\mathscr{A} = \{A: \chi_A \in M\}$ is a δ-ring and the function $A \mapsto J(\chi_A)$ is a measure, the measure induced by the integral J. Chapter 4 details this procedure and applies the result, together with the pre-measure to pre-integral to integral theorems of the preceding chapters to show that each exact function that is continuous at $\varnothing$ has an extension that is a measure. A supplement presents the standard construction of regular Borel measures and another supplement derives the existence of Haar measure.

A measure μ on a δ-ring $\mathscr{A}$ is also a pre-measure; it induces a pre-integral, and this in turn induces an integral. But there is a more direct way to obtain an integral from the measure μ: A real valued function f belongs to $L_1(\mu)$ iff there is $\{a_n\}_n$ in $\mathbb{R}$ and $\{A_n\}_n$ in $\mathscr{A}$ such that

$\sum_n |a_n| \mu(A_n) < \infty$ and $f(x) = \sum_n a_n \chi_{A_n}(x)$ for all x, and in this case the *integral* $I_\mu(f)$ is defined to be $\sum_n a_n \mu(A_n)$. This construction is given in chapter 6, and it is shown that every integral is the integral with respect to the measure it induces.

Chapter 6 requires facts about measurability that are purely set theoretic in character and these are developed in chapter 5. The critical results are: Call a function f $\mathscr{A}$ *σ-simple* (or $\mathscr{A}$ *σ^+-simple*) iff $f = \sum_n a_n \chi_{A_n}$ for some $\{A_n\}_n$ in $\mathscr{A}$ and $\{a_n\}_n$ in $\mathbb{R}$ (in $\mathbb{R}^+$, respectively). Then, if $\mathscr{A}$ is a δ-ring, a real valued function f is $\mathscr{A}$ σ-simple iff it has a support in $\mathscr{A}_\sigma$ and is locally $\mathscr{A}$ measurable (if B is an arbitrary Borel subset of $\mathbb{R}$, then $A \cap f^{-1}[B]$ belongs to $\mathscr{A}$ for each A in $\mathscr{A}$). Moreover, if such a function is non-negative, it is $\mathscr{A}$ σ^+-simple.

Chapter 7 is devoted to product measures and product integrals. It is concerned with conditions that relate the integral of a function f w.r.t. $\mu \otimes \nu$ to the iterated integrals $\int (\int f(x, y)\, d\mu x)\, d\nu y$ and $\int (\int f(x, y)\, d\nu y)\, d\mu x$. We follow the natural approach, deriving the Fubini theorem from the Tonelli theorem, and the latter leads us to grudgingly allow that some perfectly respectable σ-simple functions have infinite integrals (we call these functions *integrable in the extended sense*, or *integrable**).

Countably additive non-negative functions μ to the extended set $\mathbb{R}^*$ of reals (*measures in the extended sense* or *measures**) also arise naturally (chapter 8) as images of measures under reasonable mappings. If μ is a measure on a σ-field $\mathscr{A}$ of subsets of X, $\mathscr{B}$ is a σ-field for Y, and $T\colon X \to Y$ is $\mathscr{A} - \mathscr{B}$ measurable, then the image measure $T\mu$ is defined by $T\mu(B) = \mu(T^{-1}[B])$ for each B in $\mathscr{B}$. If $\mathscr{A}$ is a δ-ring but not a σ-field, there is a possibly infinite valued measure that can appropriately be called the T image of μ. We compute the image of Borel–Lebesgue measure for $\mathbb{R}$ under a smooth map, and so encounter indefinite integrals.

Indefinite integrals w.r.t. a σ-finite measure μ are characterized in chapter 9, and the principal result, the Radon–Nikodym theorem, is extended to decomposable measures and regular Borel measures in a supplement. Chapter 10 begins the study of Banach spaces. The duals of some standard spaces are characterized, and in a supplement our methods are used to establish very simply, or at least σ-simply, the basic facts about Bochner integrals.

This book is based on various lectures given by one or the other of us in 1965 and later, at the Indian Institute of Technology, Kanpur; Panjab University, Chandigarh; University of California, Berkeley; and the University of Kansas. We were originally motivated by curiosity about how a σ-simple approach would work; it did work, and a version of most of this text appeared as preprints in 1968, 1972 and 1979, under the title "Measures and Integrals." Since that time our point of view has changed on several matters (but not on σ-simplicity) and the techniques have been refined.

This is the first of two volumes on *Measure and Integral*. The exercises, problems, and additional supplements will appear as a companion volume to be published as soon as we can sift and edit a large disorganized mass of manuscript.

We are grateful to Klaus Bichteler, Harlan Glaz, T. Parthasarathy, and Allan Shields for suggestions and criticisms of earlier versions of this work and to Dorothy Maharam Stone and I. Namioka for their review of the final manuscript. We are indebted to our students for their comments and their insights. We owe thanks to Jean Steffey, Judy LaFollette, Carol Johnson, and especially to Ying Kelley and Sharon Gumm for assistance in preparation of the manuscript, and to Saroja Srinivasan for her nonmeasurable support.

This work was made possible by support granted at one time or another by the Miller Foundation of the University of California, Berkeley, the National Science Foundation, the Panjab (India) University, and the University of Kansas. We thank them.

J. L. KELLEY T. P. SRINIVASAN

CONTENTS

Chapter 0

PRELIMINARIES

This brief review of a few conventions, definitions and elementary propositions is for reference to be used as the need arises.

SETS

We shall be concerned with sets and with the membership relation, $\in$. If A and B are sets then $A = B$ iff A and B have the same members; i.e., for all x, $x \in A$ iff $x \in B$. A set A is a subset of a set B (B is a superset of A, $A \subset B$, $B \supset A$) iff $x \in B$ whenever $x \in A$. Thus $A = B$ iff $A \subset B$ and $B \subset A$. The empty set is denoted $\varnothing$.

If A and B are sets then the union of A and B is $A \cup B$, $\{x\colon x \in A \textit{ or } x \in B\}$; the intersection $A \cap B$ is $\{x\colon x \in A \textit{ and } x \in B\}$; the difference $A \setminus B$ is $\{x\colon x \in A \textit{ and } x \notin B$; the **symmetric difference** $A \mathbin{\triangle} B$ is $(A \cup B) \setminus (A \cap B)$; and the Cartesian product $A \times B$ is $\{(x, y)\colon x \in A, y \in B\}$. The operations of union, intersection, and symmetric difference are commutative and associative, $\cap$ distributes over $\cup$ and $\triangle$, and $\cup$ distributes over $\cap$. The set $\varnothing$ is an identity for both $\cup$ and $\triangle$.

If, for each member t of an index set T, A_t is a set, then this correspondence is called an **indexed family**, or sometimes just a **family** of sets and denoted $\{A_t\}_{t \in T}$. The union of the members of the family is $\bigcup_{t \in T} A_t = \bigcup \{A_t\colon t \in T\} = \{x\colon x \in A_t \textit{ for some member } t \textit{ of } T\}$ and the intersection is $\bigcap_{t \in T} A_t = \bigcap \{A_t\colon t \in T\} = \{x\colon x \in A_t \textit{ for each member } t \textit{ of } T\}$. There are a number of elementary identities such as $\bigcup_{t \in T \cup S} A_t = (\bigcup_{t \in T} A_t) \cup (\bigcup_{t \in S} A_t)$, $C \setminus \bigcup_{t \in T} A_t = \bigcap_{t \in T} (C \setminus A_t)$ for all sets C (the de Morgan law), and $\bigcup_{t \in T} (B \cap A_t) = B \cap \bigcup_{t \in T} A_t$.

FUNCTIONS

We write $f\colon X \to Y$, which we read as "f is on X to Y", iff f is a map of X into Y; that is, f is a function with domain X whose values belong to Y. The value of the function f at a member x of X is denoted $f(x)$, or sometimes f_x.

If $f\colon X \to Y$ then "$x \mapsto f(x)$, for x in X", is another name for f. Thus $x \mapsto x^2$, for x in $\mathbb{R}$ (the set of real numbers) is the function that sends each real number into its square. The letter "x", in "$x \mapsto x^2$ for x in $\mathbb{R}$" is a dummy variable, so $x \mapsto x^2$ for x in $\mathbb{R}$ is the same as $t \mapsto t^2$ for t in $\mathbb{R}$. (Technically, "$\mapsto$" binds the variable that precedes it.)

If $f\colon X \to Y$ and $g\colon Y \to Z$ then $\boldsymbol{g \circ f}\colon X \to Z$, the **composition** of g and f, is defined by $g \circ f(x) = g(f(x))$ for all x in X.

If $f\colon X \to Y$ and $A \subset X$ then $\boldsymbol{f|A}$ is the **restriction of f to A** (that is, $\{(x, y)\colon x \in A \text{ and } y = f(x)\}$) and $\boldsymbol{f[A]}$ is the **image of A under f** (that is, $\{y\colon y = f(x) \text{ for some } x \text{ in } A\}$). If $B \subset Y$ then $\boldsymbol{f^{-1}[B]} = \{x\colon f(x) \in B\}$ is the **pre-image** or **inverse image of B under f**. For each x, $f^{-1}[x]$ is $f^{-1}[\{x\}]$.

COUNTABILITY

A set A is **countably infinite** if there is a one to one correspondence between A and the set $\mathbb{N}$ of natural numbers (positive integers), and a set is **countable** iff it is countably infinite or finite.

Here is a list of the propositions on countability that we will use, with brief indications of proofs.

A subset of a countable set is countable.

If A is a subset of $\mathbb{N}$, define a function recursively by letting $f(n)$ be the first member of $A \setminus \{x\colon x = f(m) \text{ for some } m, m < n\}$. Then $f(n) \geqq n$ for each member n of the domain of f, and A is countably infinite if the domain of f is $\mathbb{N}$ and is finite otherwise.

The image of a countable set under a map is countable.

If f is a map of $\mathbb{N}$ onto A and $D = \{n\colon n \in \mathbb{N} \text{ and } f(m) \neq f(n) \text{ for } m < n\}$ then $f|D$ is a one to one correspondence between A and a subset of $\mathbb{N}$.

The union of a countable number of countable sets is countable.

It is straightforward to check that the union of a countable number of finite sets is countable, and $\mathbb{N} \times \mathbb{N}$ is the union, for k in $\mathbb{N}$, of the finite sets $\{(m, n)\colon m + n = k + 1\}$.

If A is an uncountable set of real numbers then for some positive integer n the set $\{a: a \in A$ and $|a| > 1/n\}$ *is uncountable.*

Otherwise A is the union of countably many countable sets.

The family of all finite subsets of a countable set is countable.

For each n in $\mathbb{N}$, the family A_n of all subsets of $\{1, \ldots\ldots, n\}$ is finite, whence $\bigcup_n A_n$ is countable.

The family of all subsets of $\mathbb{N}$ *is not countable.*

If f is a function on $\mathbb{N}$ onto the family of all subsets of $\mathbb{N}$, then for some positive integer p, $f(p) = \{n: n \notin f(n)\}$. If $p \in f(p)$ then $p \in \{n: n \notin f(n)\}$, whence $p \notin f(p)$. If $p \notin f(p)$ then $p \notin \{n: n \notin f(n)\}$, whence $p \in f(p)$. In either case there is a contradiction.

ORDERINGS AND LATTICES

A relation $\geqq$ **partially orders** a set X, or **orders** X iff it is reflexive on X ($x \geqq x$ if $x \in X$) and transitive on X (if x, y and z are in X, $x \geqq y$ and $y \geqq z$ then $x \geqq z$). A **partially ordered set** is a set X with a relation $\geqq$ that partially orders it (formally, $(x, \geqq)$ is a partially ordered set). A member u of a partially ordered set X is an **upper bound** of a subset Y of X iff $u \geqq y$ for all y in Y; and if there is an upper bound s for Y such that $u \geqq s$ for every upper bound u of Y, then s is a **supremum** of Y, *sup* Y. A **lower bound** for Y and an **infimum** of Y, *inf* Y are defined in corresponding fashion.

An ordered set X is **order complete** or **Dedekind complete** iff each non-empty subset of X that has an upper bound has a supremum, and this is the case iff each non-empty subset that has a lower bound has an infimum.

A **lattice** is a partially ordered set X such that $\{x, y\}$ has a *unique* supremum and a *unique* infimum for all x and y in X. We denote $sup\{x, y\}$ by $x \vee y$ and $inf\{x, y\}$ by $x \wedge y$. A **vector lattice** is a vector space E over the set $\mathbb{R}$ of real numbers which is a lattice under a partial ordering with the properties: for x and y in E and r in $\mathbb{R}^+$ (the set of non-negative real numbers), if $x \geqq 0$ then $rx \geqq 0$, if $x \geqq 0$ and $y \geqq 0$ then $x + y \geqq 0$, and $x \geqq y$ iff $x - y \geqq 0$. Here are some properties of vector lattices:

For all x and y, $x \vee y = -((-x) \wedge (-y))$ and $x \wedge y = -((-x) \vee (-y))$, because multiplication by -1 is order inverting.

For all x, y and z, $(x \vee y) + z = (x + z) \vee (y + z)$ *and* $(x \wedge y) + z = (x + z) \wedge (y + z)$, because the ordering is translation invariant (i.e., $x \geqq y$ iff $x + z \geqq y + z$).

For all x and y, $x + y = x \vee y + x \wedge y$ (replace z by $-x - y$ in the preceding and rearrange).

If $\boldsymbol{x}^{+} = x \vee 0$ *and* $\boldsymbol{x}^{-} = -(x \wedge 0) = (-x) \vee 0$ *then* $x = x \vee 0 + x \wedge 0 = x^{+} - x^{-}$.

For each member x of a vector lattice E, the **absolute value of** $\boldsymbol{x}$ is defined to be $|\boldsymbol{x}| = x^{+} + x^{-}$. Vectors x and y are **disjoint** iff $|x| \wedge |y| = 0$.

For each vector x, x^{+} *and* x^{-} *are disjoint,* because $x^{+} \wedge x^{-} + x \wedge 0 = (x^{+} + x \wedge 0) \wedge (x^{-} + x \wedge 0) = (x^{+} - x^{-}) \wedge 0 = x \wedge 0$, whence $x^{+} \wedge x^{-} = 0$.

The absolute value function $x \mapsto |x|$ completely characterizes the vector lattice ordering because $x \geqq 0$ iff $x = |x|$. On the other hand, if E is a vector space over $\mathbb{R}$, $A: E \to E$, $A \circ A = A$, A is **absolutely homogeneous** (i.e., $A(rx) = |r| A(x)$ for r in $\mathbb{R}$ and x in E), and A is additive on $A[E]$ (i.e., $A(A(x) + A(y)) = A(x) + A(y)$ for x and y in E), then E is a vector lattice and A is the absolute value, provided one defines $x \geqq y$ to mean $A(x - y) = x - y$.

(*Decomposition lemma*) *If* $x \geqq 0$, $y \geqq 0$, $z \geqq 0$ *and* $z \leqq x + y$, *then* $z = u + v$ *for some u and v with* $0 \leqq u \leqq x$ *and* $0 \leqq v \leqq y$. Indeed, we may set $u = z \wedge x$ and $v = z - z \wedge x$, and it is only necessary to show that $z - z \wedge x \leqq y$. But by hypothesis, $y \geqq z - x$ and $y \geqq 0$, so $y \geqq (z - x) \vee 0$, and a translation by $-z$ then shows that $y - z \geqq (-x) \vee (-z) = -(z \wedge x)$ as desired.

A real valued linear functional f on a vector lattice E is called **positive** iff $f(x) \geqq 0$ for $x \geqq 0$. If f is a positive linear functional, or if f is the difference of two positive linear functionals, then $\{f(u)\colon 0 \leqq u \leqq x\}$ is a bounded subset of $\mathbb{R}$ for each $x \geqq 0$.

If f is a linear functional on E such that $f^{+}(x) = \sup\{f(u)\colon 0 \leqq u \leqq x\} < \infty$ *for all* $x \geqq 0$, *then f is the difference of two positive linear functionals,* for the following reasons. The decomposition lemma implies that $\{f(z)\colon 0 \leqq z \leqq x + y\} = \{f(u) + f(v)\colon 0 \leqq u \leqq x \text{ and } 0 \leqq v \leqq y\}$, consequently f^{+} is additive on $\boldsymbol{P} = \{x\colon x \in E \text{ and } x \geqq 0\}$, and evidently f^{+} is absolutely homogeneous. It follows that if x, y, u and v belong to P and $x - y = u - v$, then $f^{+}(x) - f^{+}(y) = f^{+}(u) - f^{+}(v)$, and f^{+} can be extended to a linear functional on E—which we also denote by f^{+}. Moreover, $f^{+} - f$ is non-negative on P and so $f = f^{+} - (f^{+} - f)$ is the desired representation.

The class E^{*} of differences of positive linear functionals on E is itself ordered by agreeing that $f \geqq g$ iff $f(x) \geqq g(x)$ for all x in E with $x \geqq 0$. Then E^{*}, with this ordering, is a vector lattice and $f^{+} = f \vee 0$. It is to be emphasized that "f is positive" does *not* mean that $f(x) \geqq 0$ for *all* x in E, but only for members x of E with $x \geqq 0$.

Suppose a vector space F of real valued functions on a set X is ordered by agreeing that $f \geqq 0$ iff $f(x) \geqq 0$ for *all* x in X. If F, with this ordering, is a lattice, then it is a vector lattice and is called a **vector**

function lattice. This is equivalent to requiring that $(f \vee g)(x) = max\{f(x), g(x)\}$ for all x in X.

CONVERGENCE IN $\mathbb{R}^*$

A relation $\geqq$ **directs** a set D iff $\geqq$ orders D and for each α and β in D there is γ in D such that $\gamma \geqq \alpha$ and $\gamma \geqq \beta$. Examples: the usual notion of greater than or equal to directs $\mathbb{R}$, the family of finite subsets of any set X is directed by $\supset$ and also by $\subset$, and the family of infinite subsets of $\mathbb{R}$ is directed by $\supset$ but not by $\subset$.

A **net** is a pair $(x, \geqq)$ such that x is a function and $\geqq$ directs the domain D of x. We sometimes neglect to mention the order and write the net x, or the net $\{x_\alpha\}_{\alpha \in D}$. A net with values in a metric space X (or a topological space) converges to a member c of X iff $\{x_\alpha\}_{\alpha \in D}$ is *eventually* in each neighborhood U of c; that is, if for each neighborhood U of c there is α in D such that $x_\beta \in U$ for all $\beta \geqq \alpha$. If $\{x_\alpha\}_{\alpha \in D}$ converges to c and to no other point, then we write $lim_{\alpha \in D} x_\alpha = c$.

A **finite sequence** $\{x_k\}_{k=1}^n$ is a function on a set of the form $\{1, 2, \ldots, n\}$, for some n in $\mathbb{N}$. A **sequence** is a function on the set of positive integers, and the usual ordering of $\mathbb{N}$ makes each sequence a net. A sequence $\{x_n\}_{n \in \mathbb{N}}$ will also be denoted by $\{x_n\}_{n=1}^\infty$ or just by $\{x_n\}_n$. Thus for each q, $\{p + q^2\}_p$ is the sequence $p \mapsto p + q^2$ for p in $\mathbb{N}$.

It is convenient to extend the system of real numbers. The set $\mathbb{R}$, with two elements ∞ and $-\infty$ adjoined, is the **extended set** $\mathbb{R}^*$ of real numbers and members of $\mathbb{R}^*$ are **real*** numbers. We agree that ∞ is the largest member of $\mathbb{R}^*$, $-\infty$ is the smallest, and for each r in $\mathbb{R}$ we agree that $r + \infty = \infty + r = \infty, r + -\infty = -\infty + r = -\infty, r \cdot \infty = \infty$ if $r > 0$, $r \cdot \infty = -\infty$ if $r < 0$, $r \cdot (-\infty) = (-r) \cdot \infty$ for $r \neq 0$, $0 \cdot \infty = 0 \cdot (-\infty) = 0$, $\infty \cdot \infty = (-\infty) \cdot (-\infty) = \infty$ and $\infty \cdot (-\infty) = (-\infty) \cdot \infty = -\infty$.

Every non-empty subset of $\mathbb{R}$ which has an upper bound has a smallest upper bound, or supremum, in $\mathbb{R}$ and it follows easily that every subset of $\mathbb{R}^*$ has a supremum in $\mathbb{R}^*$ and also an infimum. In particular, $sup\, \varnothing = -\infty$ and $inf\, \varnothing = -\infty$.

A **neighborhood** in $\mathbb{R}^*$ of a member r of $\mathbb{R}$ is a subset of $\mathbb{R}^*$ containing an open interval about r. A subset V of $\mathbb{R}^*$ is a **neighborhood** of ∞ iff for some real number r, V contains $\{s \colon s \in \mathbb{R}^* \text{ and } s > r\}$. Neighborhoods of $-\infty$ are defined in a corresponding way. Consequently a net $\{x_\alpha\}_{\alpha \in D}$ in R^* converges to ∞ iff for each real number s there is β in D such that $x_\alpha > s$ for $\alpha \geqq \beta$.

If $\{x_\alpha\}_{\alpha \in A}$ and $\{y_\alpha\}_{\alpha \in A}$ are convergent nets in R^* then $lim_{\alpha \in A}(x_\alpha + y_\alpha) = lim_{\alpha \in A} x_\alpha + lim_{\alpha \in A} y_\alpha$, provided the sum of the limits is defined and $lim_{\alpha \in A} x_\alpha y_\alpha = (lim_{\alpha \in A} x_\alpha) \cdot (lim_{\alpha \in A} y_\alpha)$ provided the pair $(lim_{\alpha \in A} x_\alpha, lim_{\alpha \in A} y_\alpha)$ is not one among $(0, \pm\infty)$ or $(\pm\infty, 0)$. The proofs parallel those for nets in $\mathbb{R}$ with minor modifications.

If a net $\{x_\alpha\}_{\alpha \in A}$ in $\mathbb{R}^*$ is **increasing** (more precisely **non-decreasing**) in the sense that $x_\beta \geqq x_\alpha$ if $\beta \geqq \alpha$, then $\{x_\alpha\}_{\alpha \in A}$ converges to $sup_{\alpha \in A} x_\alpha$; for if $r < sup_{\alpha \in A} x_\alpha$, then r is not an upper bound for $\{x_\alpha\}_{\alpha \in A}$, consequently $r < x_\alpha$ for some α, and hence $r < x_\beta \leqq sup_{\alpha \in A} x_\alpha$ for $\beta \geqq \alpha$. Likewise, a decreasing net in R^* converges to $inf_{\alpha \in A} x_\alpha$ in R^*.

If $\{x_\alpha\}_{\alpha \in A}$ is a net in $\mathbb{R}^*$ then $\alpha \mapsto sup\{x_\beta: \beta \in A \text{ and } \beta \geqq \alpha\}$ is a decreasing net and consequently converges to a member of $\mathbb{R}^*$. This member is denoted $\mathbf{lim\,sup}_{\alpha \in A} x_\alpha$ or $\mathbf{lim\,sup}\{x_\alpha: \alpha \in A\}$. Similarly $\mathbf{lim\,inf}\{x_\alpha: \alpha \in A\}$ is $lim_{\alpha \in A} inf\{x_\beta: \beta \geqq \alpha\}$. It is easy to check that a net $\{x_\alpha\}_{\alpha \in A}$ converges iff $lim\,sup_{\alpha \in A} x_\alpha = lim\,inf_{\alpha \in A} x_\alpha$, and that in this case $lim_{\alpha \in A} x_\alpha = lim\,sup_{\alpha \in A} x_\alpha = lim\,inf_{\alpha \in A} x_\alpha$.

If $\{f_\alpha\}_{\alpha \in A}$ is a net of functions on a set X to $\mathbb{R}^*$ then $\mathbf{sup}_{\alpha \in A} f_\alpha$ is defined to be the function whose value at x is $sup_{\alpha \in A} f_\alpha(x)$, and similarly, $(\mathbf{inf}_{\alpha \in A} f_\alpha)(x) = inf_{\alpha \in A} f_\alpha(x)$, $(\mathbf{lim\,sup}_{\alpha \in A} f_\alpha)(x) = lim\,sup_{\alpha \in A} f_\alpha(x)$ and $(\mathbf{lim\,inf}_{\alpha \in A} f_\alpha)(x) = lim\,inf_{\alpha \in A} f_\alpha(x)$. The net $\{f_\alpha\}_{\alpha \in A}$ **converges pointwise** to f iff $f = lim\,sup_{\alpha \in A} f_\alpha = lim\,inf_{\alpha \in A} f_\alpha$ or, equivalently, $f(x) = lim_{\alpha \in A} f_\alpha(x)$ for all x.

UNORDERED SUMMABILITY

Suppose $x = \{x_t\}_{t \in T}$ is an indexed family of real* numbers. We agree that $\{x_t\}_{t \in T}$ is **summable* over a finite subset** A of T iff x does not assume both of the values ∞ and $-\infty$ at members of A, and in this case the sum of x_t for t in A is denoted by $\sum_{t \in A} x_t$ or $\sum_A \boldsymbol{x}$. If $\{x_t\}_{t \in T}$ is summable* over each finite subset, and if $\mathscr{F}$ is the class of all finite subsets of T, then $\mathscr{F}$ is directed by $\supset$, $\{\sum_A x\}_{A \in \mathscr{F}}$ is a net, and we say that x is **summable* over T**, or just **summable*** provided that the net $\{\sum_A x\}_{A \in \mathscr{F}}$ converges. In this case the **unordered sum**, $\sum_T x$, is $lim\{\sum_A x: A \in \mathscr{F}\}$, and $\{x_t\}_{t \in T}$ is summable* to $\sum_T x$.

If $x = \{x_t\}_{t \in T}$ is a family of real numbers, then x is automatically summable* over each finite subset of T and we say that x is **summable over T**, or just **summable**, provided it is summable* and $\sum_T x \in \mathbb{R}$.

If $\{x_n\}_{n \in \mathbb{N}}$ is a sequence of real numbers then the (ordered) sum, $lim_n \sum_{k=1}^n x_k$, may exist although the sequence is not summable (e.g., $x_n = (-1)^n/n$ for each n in $\mathbb{N}$). However, if $\{x_n\}_n$ is summable* then the limit of $\{\sum_{k=1}^n x_k\}_n$ exists and $lim_n \sum_{k=1}^n x_k = \sum_{n \in \mathbb{N}} x_n$.

Here are the principal facts about unordered summation, with a few indications of proof. Throughout, $x = \{x_t\}_{t \in T}$ and $y = \{y_t\}_{t \in T}$ will be indexed families of real* numbers, $(x^+)_t = (x_t)^+$ and $(x^-)_t = (x_t)^-$ for each t, and r will be a real number.

The family $x = \{x_t\}_{t \in T}$ is summable iff for $e > 0$ there is a finite subset A of T such that $\sum_B |x| < e$ for each finite subset B of $T \setminus A$.

If $x = \{x_t\}_{t \in T}$ is summable then $x_t = 0$ except for countably many points t.

If $x_t \geqq 0$ for each t then $\{x_t\}_{t \in T}$ is summable.*

(The net $\{\sum_A x\colon A \in \mathscr{F}\}$ is increasing.)

The family x is summable iff one of $\sum_T x^+$ and $\sum_T x^-$ is finite; it is summable iff both are finite; and in either of these two cases, $\sum_T x = \sum_T x^+ - \sum_T x^-$.* (The result reduces to the usual "limit of the difference" proposition.)

If x is summable and $r \in \mathbb{R}$ then rx is summable* and $\sum_T rx = r\sum_T x$.*

The next proposition states that "$\sum_X$ is additive except for $\infty - \infty$ troubles". It's another "limit of a sum" result.

If x and y are summable, $\{x_t, y_t\} \neq \{\infty, -\infty\}$ for all t, and $\{\sum_T x, \sum_T y\} \neq \{\infty, -\infty\}$, then $x + y$ is summable* and $\sum_T (x + y) = \sum_T x + \sum_T y$.*

If x is summable over T and $A \subset T$ then x is summable* over A.*

If x is summable over T and $\mathscr{B}$ is a disjoint finite family of subsets of T then $\sum_{B \in \mathscr{B}} (\sum_B x) = \sum \{x_t\colon t \in \bigcup_{B \in \mathscr{B}} B\}$.*

If $\mathscr{A}$ is a decomposition of T (i.e., a disjoint family of subsets such that $T = \bigcup_{A \in \mathscr{A}} A$) and x is summable over T then $A \mapsto \sum_A x$ is summable* over $\mathscr{A}$ and $\sum_T x = \sum_{A \in \mathscr{A}} \sum_A x$.*

If x is summable over $Y \times Z$, then $\sum_{Y \times Z} x = \sum_{y \in Y} \sum_{z \in Z} x(y, z) = \sum_{z \in Z} \sum_{y \in Y} x(y, z)$.*

It is worth noticing that the condition, "*x is summable**", is necessary for the last equality. Here is an example. Define x on $\mathbb{N} \times \mathbb{N}$ by letting $x(m, n)$ be 1 if $m = n$, -1 if $n = m + 1$, and 0 otherwise. Then $\sum_{m \in \mathbb{N}} x(m, n) = 0$ if $n > 0$ and 1 if $n = 0$, so $\sum_{n \in \mathbb{N}} (\sum_{m \in \mathbb{N}} x(m, n)) = 1$, whereas $\sum_{m \in \mathbb{N}} (\sum_{n \in \mathbb{N}} x(m, n)) = \sum_{m \in \mathbb{N}} (0) = 0$.

A family $\{f_t\}_{t \in T}$ of real* valued functions on a set X is **pointwise summable*** (**summable**, respectively) iff $\{f_t(x)\}_{t \in T}$ is summable* (summable, respectively) for each x in X, and in this case the **pointwise sum**, $(\sum_{t \in T} f_t)(x)$ is defined to be $\sum_{t \in T} f_t(x)$ for each x in X.

HAUSDORFF MAXIMAL PRINCIPLE

If $\geqq$ partially orders X then a subset C of X is a **chain** iff for all x and y in C with $x \neq y$, either $x \geqq y$ or $y \geqq x$ but not both. We assume (and occasionally use) the following form of the maximal principle.

ZORN'S LEMMA *If C is a chain in a partially ordered space $(X, \geqq)$ then C is contained in a maximal chain D—that is a chain that is a proper subset of no other chain.*

Consequently, if every chain in X has a supremum in X then there is a maximal member m of X—that is, *if $n \geqq m$ then $n = m$.*

Here is a simple example of the application of the maximal principle. Suppose that G is a subset of the real plane $\mathbb{R}^2$ and that $\mathscr{D}$ is the family of disks $D_r(a, b) = \{(x, y)\colon (x - a)^2 + (y - b)^2 \leqq r^2\}$ with (a, b) in $\mathbb{R}^2$, $r > 0$ and $D_r(a, b) \subset G$. Then there is a maximal disjoint subfamily $\mathscr{M}$ of $\mathscr{D}$, and $G \setminus \bigcup_{D \in \mathscr{M}} D$ contains no non-empty open set.

Chapter 1

PRE-MEASURES

We consider briefly the class of length functions. These will turn out to be precisely the functions on the family of closed intervals that can be extended to become measures; these are examples of pre-measures. Their theory furnishes a concrete illustration of the general construction of measures.

A **closed interval** is a set of the form $\mathbf{[a:b]} = \{x\colon x \in \mathbb{R} \text{ and } a \leqq x \leqq b\}$, an **open interval** is a set of the form $(a:b) = \{x\colon a < x < b\}$, and $(a:b]$ and $[a:b)$ are half open intervals. The family of closed intervals is denoted $\mathscr{J}$; we agree that $\varnothing \in \mathscr{J}$. We are concerned with real valued functions λ on $\mathscr{J}$, and we abbreviate $\lambda([a:b])$ by $\boldsymbol{\lambda[a:b]}$. The closed interval $[b:b]$ is just the singleton $\{b\}$, and $\lambda[b:b] = \lambda(\{b\})$ is abbreviated $\boldsymbol{\lambda\{b\}}$.

A non-negative real valued function λ on $\mathscr{J}$ such that $\lambda(\varnothing) = 0$ is a **length**, or a **length function for** $\mathbb{R}$, iff λ has three properties:

Boundary inequality If $a < b$ then $\lambda[a:b] \geqq \lambda\{a\} + \lambda\{b\}$.
Regularity If $a \in \mathbb{R}$ then $\lambda\{a\} = \inf\{\lambda[a - e:a + e]:e > 0$.
Additive property If $a \leqq b \leqq c$ then $\lambda[a:b] + \lambda[b:c] = \lambda[a:c] + \lambda[b:b]$.

The **length**, or the **usual length function** ℓ, is defined by $\ell[a:b] = b - a$ for $a \leqq b$. The length ℓ is evidently a length function; it has a number of special properties—for example, $\lambda\{x\} = 0$ for all x.

There are length functions that vanish except at a singleton. The **unit mass at a member** $\boldsymbol{c}$ of $\mathbb{R}$, ε_c, is defined by letting $\varepsilon_c[a:b]$ be one if $c \in [a:b]$ and zero otherwise. Thus $\varepsilon_c\{x\} = 0$ if $x \neq c$ and $\varepsilon_c\{c\} = 1$. Each such unit mass is a length function, and each non-negative, finite linear combination of unit masses is a length function.

A length function λ is **discrete** iff $\lambda[a:b] = \sum_{x \in [a:b]} \lambda\{x\}$ for every closed interval $[a:b]$. That is, a length function λ is discrete iff the function $x \mapsto \lambda\{x\}$ is summable over each closed interval $[a:b]$ and $\lambda[a:b]$ is the sum $\sum_{x \in [a:b]} \lambda\{x\}$ (of course, in this case $\lambda\{x\} = 0$ except for countably many x). Each discrete length λ is the sum $\sum_{x \in \mathbb{R}} \lambda\{x\} \varepsilon_x$, since $\sum_{x \in \mathbb{R}} \lambda\{x\} \varepsilon_x[a:b] = \sum_{x \in [a:b]} \lambda\{x\} = \lambda[a:b]$.

If λ is a discrete length function then the function $x \mapsto \lambda\{x\}$ determines λ entirely. On the other hand, if f is a non-negative real valued function that is summable over intervals and $\lambda[a:b] = \sum_{x \in [a:b]} f(x)$, then λ evidently satisfies the boundary inequality and has the additive property required for length functions. It is also regular, and hence a discrete length function, as the following argument shows. If $a \in \mathbb{R}$, $e > 0$ and $E = [a - 1:a + 1] \setminus \{a\}$, then there is a finite subset F of E such that $\sum_{x \in E} f(x) < e + \sum_{x \in F} f(x)$, whence $\sum_{x \in E \setminus F} f(x) < e$. If $d < min\{|x - a| : x \in F\}$, then $\sum_{x \in [a-d:a+d]} f(x) \leqq f(a) + \sum_{x \in E \setminus F} f(x) < f(a) + e$. Thus $\lambda[a - d:a + d] < \lambda\{a\} + e$, and consequently $\lambda\{a\} = inf\{\lambda[a - d:a + d]:d > 0\}$.

A length function λ is **continuous** iff $\lambda\{x\} = 0$ for all x. The usual length function ℓ is continuous. Another example of a continuous length function: if f is a non-negative real valued continuous function on $\mathbb{R}$ and $\lambda[a:b]$ is the Riemann integral of f over $[a:b]$, then λ is a continuous length function.

It turns out that each length function is the sum, in a unique way, of a discrete length function and a continuous one. We prove this after establishing a lemma.

1 LEMMA *If λ is a length function and $a = a_0 \leqq a_1 \leqq \cdots \leqq a_{m+1} = b$, then $\sum_{i=0}^{m} \lambda[a_i:a_{i+1}] = \lambda[a:b] + \sum_{i=1}^{m} \lambda\{a_i\}$, and if $a_i < a_{i+1}$ for each i, then $\lambda[a:b] \geqq \sum_{i=0}^{m+1} \lambda\{a_i\}$.*

PROOF The definition of length implies the lemma for $m = 1$. Assume that the proposition is established for $m = p$ and that $a_0 \leqq a_1 \leqq \cdots \leqq a_{p+2}$. Then $\sum_{i=0}^{p} \lambda[a_i:a_{i+1}] = \lambda[a_0:a_{p+1}] + \sum_{i=1}^{p} \lambda\{a_i\}$, hence $\sum_{i=0}^{p+1} \lambda[a_i:a_{i+1}] = \lambda[a_0:a_{p+1}] + \lambda[a_{p+1}:a_{p+2}] + \sum_{i=1}^{p} \lambda\{a_i\}$, and the additivity property of λ then implies that $\sum_{i=0}^{p+1} \lambda[a_i:a_{i+1}] = \lambda[a_0:a_{p+2}] + \sum_{i=1}^{p+1} \lambda\{a_i\}$.

If $a_i < a_{i+1}$ for each i, then the boundary inequality implies that $\sum_{i=0}^{m} \lambda[a_i:a_{i+1}] \geqq \sum_{i=0}^{m} (\lambda\{a_i\} + \lambda\{a_{i+1}\})$, so $\lambda[a:b] + \sum_{i=1}^{m} \lambda\{a_i\} \geqq \sum_{i=0}^{m} \lambda\{a_i\} + \sum_{i=1}^{m+1} \lambda\{a_i\}$ and hence $\lambda[a:b] \geqq \sum_{i=0}^{m+1} \lambda\{a_i\}$. ∎

It is a consequence of the preceding that each length function is **monotonic**; that is, if $[c:d] \subset [a:b]$ then $\lambda[c:d] \leqq \lambda[a:b]$. If $a < c < d < b$ then $\lambda[a:c] + \lambda[c:d] + \lambda[d:b] = \lambda[a:b] + \lambda\{c\} + \lambda\{d\}$, so $\lambda[a:b] - \lambda[c:d] = \lambda[a:c] - \lambda\{c\} + \lambda[d:b] - \lambda\{d\} \geqq 0$, and the various special cases (e.g., $a = c$) are easy to check.

Suppose λ is a length function. The **discrete part** of λ, λ_d is defined by

$\lambda_d(I) = \sum_{x \in I} \lambda\{x\}$ for each closed interval I. The inequality asserted in the preceding lemma states that $\lambda_d(I) \leqq \lambda(I)$ for each I in $\mathscr{I}$, and it follows that $x \mapsto \lambda\{x\}$ is summable over each interval, and consequently λ_d is a length function. It is a discrete length function because $\lambda_d[a:b] = \sum_{x \in [a:b]} \lambda\{x\} = \sum_{x \in [a:b]} \lambda_d\{x\}$.

The **continuous part** λ_c of the length function λ is defined by $\lambda_c(I) = \lambda(I) - \lambda_d(I)$ for all closed intervals I. The function λ_c is non-negative because $\lambda_d \leqq \lambda$, and it is straightforward to check that it satisfies the boundary inequality and has the additive property for length. Finally, $\lambda_c\{x\} = \lambda\{x\} - \lambda_d\{x\} = 0$ for all x, and $\mathit{inf}\{\lambda_c[x - e:x + e]:e > 0\} = \mathit{inf}\{\lambda[x - e:x + e] - \lambda_d[x - e:x + e]:e > 0\} = 0$ because λ is regular, so λ_c has the regularity property, and consequently it is a continuous length.

We have seen that each length function λ can be represented as the sum $\lambda_c + \lambda_d$ of a continuous length and a discrete length. The representation is in fact unique, for if $\lambda = \lambda_1 + \lambda_2$ where λ_1 is a discrete length and λ_2 is continuous then $\lambda\{x\} = \lambda_1\{x\} + \lambda_2\{x\} = \lambda_1\{x\}$ because λ_2 is continuous, and since λ_1 is discrete, $\lambda_1(I) = \sum_{x \in I} \lambda_1\{x\} = \sum_{x \in I} \lambda\{x\} = \lambda_d(I)$ for all closed intervals I. Consequently $\lambda_1 = \lambda_d$ and $\lambda_c = \lambda_2$.

We record this result for reference.

2 PROPOSITION *Each length function is the sum in just one way of a discrete length and a continuous length.*

There is a standard way of manufacturing length functions. Suppose f is a real valued function on $\mathbb{R}$ that is **increasing** in the sense that $f(x) \geqq f(y)$ whenever $x \geqq y$. For each x in $\mathbb{R}$ let $f_-(x)$, the **left hand limit** of f at x, be $\mathit{sup}\{f(y): y < x\}$ and let $f_+(x)$, the **right hand limit** of f at x, be $\mathit{inf}\{f(y): y > x\}$. It is easy to verify that f_+ is increasing and **right continuous** (that is, $(f_+)_+ = f_+$) and that f_- is increasing and **left continuous.** The **jump of f at x, $j_f(x)$,** is $f_+(x) - f_-(x) = \mathit{inf}\{f(x + e) - f(x - e):e > 0\}$; it is 0 iff f is continuous at x. The function f is called a **jump function** provided $f_+(b) - f_-(a) = \sum_{x \in [a:b]} j_f(x)$ for all a and b with $a \leqq b$.

The **f length λ_f,** or the **length induced by f,** is defined by $\lambda_f[a:b] = f_+(b) - f_-(a)$ for all a and b with $a \leqq b$. We note that $\lambda_f\{x\}$ is just the jump, $j_f(x)$.

3 PROPOSITION *If f is an increasing function on $\mathbb{R}$ to $\mathbb{R}$ then λ_f is a length function; it is a continuous length iff f is continuous and is discrete iff f is a jump function.*

PROOF A straightforward verification shows that λ_f satisfies the boundary inequality and has the additive property for length. If $b \in \mathbb{R}$

and $e > 0$ then $inf\{\lambda_f[b - e:b + e]:e > 0\} = inf\{f_+(b + e):e > 0\} - sup(f_-(b - e):e > 0\}$. But f_+ is right continuous and f_- is left continuous, hence $inf\{\lambda_f[b + e:b - e]:e > 0\} = f_+(b) - f_-(b) = \lambda_f\{b\}$, so λ_f is regular and hence is a length function.

The length λ_f is continuous iff $\lambda_f\{x\} = j_f(x) = 0$ for all x; that is, f is a continuous function. The function λ_f is discrete iff $\lambda_f[a:b] = \sum_{x \in [a:b]} \lambda_f\{x\}$ and this is the case iff $f_+(b) - f_-(a) = \sum_{x \in [a:b]} j_f(x)$; that is, if f is a jump function. ■

We will show that every length function is f length for some f. It will then follow from propositions 2 and 3 that each increasing f is in just one way the sum of a jump function and a continuous function.

Different increasing functions F may induce the same length, and in particular F, F + (a constant), F_+, F_- and any function sandwiched between F_- and F_+ all induce the same length. We agree that F is a **distribution function for a length** λ iff $\lambda = \lambda_F$. **A normalized distribution function** for a length λ is a right continuous increasing function F that induces λ and vanishes at 0 (one could, alternatively, "normalize" by pre-assigning a different value or a value at a different point and/or require left continuity in place of right).

4 PROPOSITION *The unique normalized distribution function F for a length λ is given by $F(x) = \lambda[0:x] - \lambda\{0\}$ for $x \geqq 0$ and $F(x) = -\lambda[x:0] + \lambda\{x\}$ for $x < 0$; alternatively, $F(x) = \lambda[a:x] - \lambda[a:0]$ for each x and all $a \leqq \min\{x, 0\}$.*

PROOF If $a \leqq b \leqq c$ then $\lambda[a:c] - \lambda[a:b] = \lambda[b:c] - \lambda\{b\}$ by the additive property. It follows that if $a \leqq x$, $a \leqq 0$ and $F(x) = \lambda[a:x] - \lambda[a:0]$ then $F(x)$ does not depend on a, and that $F(x) = \lambda[0:x] - \lambda\{0\}$ for $x \geqq 0$ and $F(x) = -\lambda[x:0] + \lambda\{x\}$ for $x < 0$. Evidently $F(0) = 0$, and if $e > 0$, $a \leqq x$ and $a \leqq 0$ then $F(x + e) - F(x) = \lambda[a:x + e] - \lambda[a:x] = \lambda[x:x + e] - \lambda\{x\}$, so right continuity of F is a consequence of the regularity of λ.

If $b \leqq c$ and $a \leqq min\{b, 0\}$, then $F(c) - F(b) = \lambda[a:c] - \lambda[a:0] - (\lambda[a:b] - \lambda[a:0]) = \lambda[a:c] - \lambda[a:b] = \lambda[b:c] - \lambda\{b\}$. If we show that $F(b) = F_-(b) + \lambda\{b\}$, then it will follow that $F(c) - F_-(b) = \lambda[b:c]$ for all $b \leqq c$, whence F is a distribution function for λ. For $a < b$, $F(b) - F(a) = \lambda[a:b] - \lambda\{a\}$ and if a is near b, then $\lambda[a, b]$ is near $\lambda\{b\}$ by regularity. Moreover, since $a \mapsto \lambda\{a\}$ is summable over each interval,$\{\lambda\{a_n\}\}_n$ converges to zero for each strictly increasing sequence $\{a_n\}_n$ that converges to b. Hence $F(b) - F_-(b) = \lambda\{b\}$, and it follows that F is a normalized distribution function for λ.

Finally, if G is also a normalized distribution function for λ then $F(x) - F_-(a) = \lambda[a:x] = G(x) - G_-(a)$ for $a \leqq x$ so F and G differ by a constant, and since $F(0) = G(0) = 0$ this constant is zero. ■

The usual length function ℓ, where $\ell[a:b] = b - a$ for $a \leqq b$, is characterized among length functions λ by the fact that for $\lambda = \ell$, $\lambda[0:1] = 1$ and λ is **invariant under translation**, in the sense that $\lambda[a:b] = \lambda[a + x:b + x]$ for all x and all a and b with $a \leqq b$. If we agree that the **translate of a set E by x, $E + x$,** is $\{y + x : y \in E\}$ then $\ell(E + x) = \ell(E)$ for each E in $\mathscr{J}$.

5 THEOREM *There is, to a constant multiple, a unique translation invariant length—each invariant length λ is $\lambda[0:1]\ell$.*

PROOF Suppose λ is a translation invariant length. Then $\lambda\{x\} = \lambda\{y\}$ for all x and y in $\mathbb{R}$ because $y = x + (x - y)$, and since $\infty > \lambda[0:1] \geqq \sum_{x \in [0:1]} \lambda\{x\}$, it must be that $\lambda\{x\} = 0$ for all x. Thus λ is a continuous length so $\lambda[a:b] + \lambda[b:c] = \lambda[a:c]$ for $a \leqq b \leqq c$. Moreover, $\lambda[b:c] = \lambda[0:c - b]$ for $b \leqq c$ because λ is translation invariant.

Let $f(x) = \lambda[0:x]$ for $x \geqq 0$. Then f is monotonic and for x and y non-negative, $f(x + y) = \lambda[0:x + y] = \lambda[0:x] + \lambda[x:x + y] = f(x) + f(y)$. Consequently, by induction, $f(nx) = nf(x)$ for n in $\mathbb{N}$ and $x \geqq 0$, and letting $y = x/n$, we infer that $f(y/n) = (1/n)f(y)$. Therefore $f(rx) = rf(x)$ for all $x \geqq 0$ and all rational non-negative r, and so $f(r) = rf(1)$. Finally, f is monotonic, so $sup\{f(r):r$ rational and $r \leqq x\} \leqq f(x) \leqq inf\{f(r):r$ rational and $r \geqq x\}$, whence $xf(1) = sup\{rf(1):r$ rational and $r \leqq x\} \leqq f(x) \leqq inf\{rf(1):r$ rational and $r \geqq x\} = xf(1)$, so $f(x) = xf(1)$ for $x \geqq 0$. Thus $\lambda[b:c] = f(c - b) = (c - b)f(1) = \ell[b:c]\lambda[0:1]$ for $b \leqq c$. ■

We shall eventually extend each length function λ to a domain substantially larger than the family $\mathscr{J}$ of closed intervals. We begin by extending λ to the class of unions of finitely many closed intervals.

A **lattice** of sets is a non-empty family $\mathscr{A}$ that is closed under finite union and intersection. That is, a non-empty family $\mathscr{A}$ is a lattice iff $A \cup B$ and $A \cap B$ belong to $\mathscr{A}$ for all members A and B of $\mathscr{A}$. The inclusion relation partially orders each family $\mathscr{A}$, and $\mathscr{A}$ is a lattice with this partial ordering iff $\mathscr{A}$ is a lattice of sets. The family of all finite subsets of $\mathbb{R}$, or of all countable subsets, or of all compact subsets or of all open subsets, are examples of lattices.

The lattice **$\mathscr{L}(\mathscr{A})$ generated by a family** $\mathscr{A}$ of sets is the smallest lattice of sets that contains $\mathscr{A}$. Evidently $\mathscr{L}(\mathscr{A})$ consists of finite unions of finite intersections of members of $\mathscr{A}$. The family $\mathscr{J}$ of closed intervals is closed under finite intersection and the union of two intersecting intervals is an interval, so $\mathscr{L}(\mathscr{J})$ is the class of unions of finitely many disjoint closed intervals.

An **exact function** is a real valued non-negative function μ on a lattice $\mathscr{A}$ such that: $\varnothing \in \mathscr{A}$, $\mu(\varnothing) = 0$, and $\mu(A) = \mu(B) + sup\{\mu(C): C \in \mathscr{A}$ and $C \subset A \setminus B$ for all A and B in $\mathscr{A}$ with $B \subset A\}$. An exact function μ is automatically monotonic (if $A \supset B$ then $\mu(A) \geqq \mu(B)$), and exactness also implies that $\mu(A \cup B) = \mu(A) + \mu(B)$ for all disjoint members A and B of with $A \cup B$ in $\mathscr{A}$ (that is, μ is **additive**).

We show that each length function λ on $\mathscr{J}$ has a unique extension (its **canonical extension**) to an exact function on $\mathscr{L}(\mathscr{J})$.

6 THEOREM *Each length function λ on $\mathscr{J}$ extends uniquely to an exact function μ on the lattice $\mathscr{L}(\mathscr{J})$ of unions of finitely many closed intervals.*

PROOF The only possible exact extension of a length function λ to $\mathscr{L}(\mathscr{J})$ is given by $\mu(\bigcup_{i=1}^{m} I_i) = \sum_{i=1}^{m} \lambda(I_i)$ for each disjoint family $\{I_i\}_{i=1}^{m}$ with I_i in $\mathscr{J}$, so the proof reduces to showing μ is exact. For convenience, let $\lambda_*(E) = sup\{\lambda(I): I \subset E$ and $I \in \mathscr{J}\}$ and let $\mu_*(E) = sup\{\mu(D): D \subset E$ and $D \in \mathscr{L}(\mathscr{J})\}$ for $E \subset \mathbb{R}$. It is straightforward to verify, using the definition of length function, that if $a < b$ then $\lambda_*(a:b) = \lambda[a:b] - \lambda\{a\} - \lambda\{b\}$, $\lambda_*[a:b) = \lambda[a:b] - \lambda\{b\}$ and $\lambda_*(a:b] = \lambda[a:b] - \lambda\{a\}$.

Suppose $c_1 \leq d_1 < c_2 \leq d_2 < \cdots < c_n \leq d_n$. Then by lemma 1, $\lambda[c_1:d_n] = \sum_{i=1}^{n} \lambda[c_i:d_i] + \sum_{i=1}^{n-1} (\lambda[d_i:c_{i+1}] - \lambda\{d_i\} - \lambda\{c_{i+1}\}) = \sum_{i=1}^{n} \lambda[c_i:d_i] + \sum_{i=1}^{n-1} \lambda_*(d_i:c_{i+1}) \geq \sum_{i=1}^{n} \lambda[c_i:d_i]$. If E is an interval— open, closed or half-open—and $E \supset \bigcup_{i=1}^{n} [c_i:d_i]$ then $E \supset [c_1:d_n]$ and it follows that $\mu_*(E) = \lambda_*(E)$.

If $A = [a:b] \supset B = \bigcup_{i=1}^{n} [c_i:d_i]$ then $\mu(A) = \lambda(A) = \lambda[a:c_1] - \lambda\{c_1\} + \lambda[c_1:d_n] + \lambda[d_n:b] - \lambda\{d_n\} = \mu_*[a:c_1) + \mu(B) + \sum_{i=1}^{n-1} \mu_*(d_i:c_{i+1}) + \mu_*(d_n:b]$. If E and F are intervals and $sup E < inf F$ then $\mu_*(E \cup F) = \mu_*(E) + \mu_*(F)$. It follows that $\mu(A) = \mu(B) + \mu_*(A \setminus B)$. Finally, this last equality extends without difficulty to a union A of finitely many disjoint closed intervals. ■

SUPPLEMENT: CONTENTS

The extended length function of theorem 6 is a special case of a more general construct. Let us suppose that X is a locally compact Hausdorff space. A **content** for X is a non-negative real valued, subadditive, additive, monotonic function μ on the family $\mathscr{C}$ of compact sets. That is, for all A and B in $\mathscr{C}$, $0 \leq \mu(A) < \infty$, $\mu(A \cup B) \leq \mu(A) + \mu(B)$ with equality if $A \cap B = \varnothing$, and $\mu(A) \leq \mu(B)$ if $A \subset B$. A content μ is **regular** iff for each member A of $\mathscr{C}$ and each $e > 0$ there is a member B of $\mathscr{C}$ with A a subset of the **interior** B^0 of B and $\mu(B) - \mu(A) < e$. Thus, μ is regular iff $\mu(A) = inf\{\mu(B): B \in \mathscr{C}$ and $B^0 \supset A\}$. A content may fail to be regular but each content can be "regularized" in the following sense. The **regularization** μ' of a content μ is defined by $\mu'(A) = inf\{\mu(B): A \subset B^0, B$ compact$\}$ for all compact sets A.

7 PROPOSITION *The regularization μ' of a content μ is a regular content.*

PROOF It is easy to see that μ' is regular; we have to show that it is a content.

Clearly μ' is monotone, non-negative and real valued. Suppose that A and B are compact and C and D are members of $\mathscr{C}$, that $A \subset C^0$ and $B \subset D^0$. Then $A \cup B \subset (C \cup D)^0$ and hence $\mu'(A \cup B) \leqq \mu(C \cup D) \leqq \mu(C) + \mu(D)$. Taking the infimum for all such C and D, we see that $\mu'(A \cup B) \leqq \mu'(A) + \mu'(B)$, so μ' is subadditive.

It remains to prove that μ' is additive. Suppose that A and B are disjoint compact sets and that $A \cup B \subset C^0$ where $C \in \mathscr{C}$. Then we may choose members E and F of $\mathscr{C}$ so that $E \cap F = \varnothing$, $E^0 \supset A$, $F^0 \supset B$ and $E \cup F \subset C$. Then $\mu(C) \geqq \mu(E \cup F) = \mu(E) + \mu(F) \geqq \mu'(A) + \mu'(B)$. Taking the infimum for all such C shows that $\mu'(A \cup B) \geqq \mu'(A) + \mu'(B)$. ■

There is a variant of the preceding that is sometimes useful. Let us agree that a **pre-content** for X is a non-negative, real valued, subadditive, additive, monotonic function μ on a class $\mathscr{B}$ of compact subsets of X with the properties: the union of two members of $\mathscr{B}$ belongs to $\mathscr{B}$, and $\mathscr{B}$ is a **base for neighborhoods of compacta** in the sense that every neighborhood of a compact set A contains a compact neighborhood of A that belongs to $\mathscr{B}$. The pre-content μ is **regular** iff its **regularization** μ', given by $\mu'(A) = inf\{\mu(B): B \in \mathscr{B}$ and $A \subset B^0$ for compact A, agrees with μ on $\mathscr{B}$.

The argument for the preceding proposition shows that the regularization of a pre-content μ on $\mathscr{B}$ is a regular content μ'. If a regular content v is an extension of a pre-content μ on $\mathscr{B}$, then $v = \mu'$, for the following reasons. Each compact neighborhood A of a compact set C contains a compact neighborhood B that belongs to $\mathscr{B}$, so $C \subset B \subset A$ and $v(C) \leqq v(B) = \mu(B) \leqq v(A)$. Hence $v(C) \leqq \mu'(C) \leqq v(A)$, and since v is regular, $v = \mu'$. Thus:

8 PROPOSITION *The regularization of a pre-content μ is a regular content μ', and if μ is regular, then μ' is the unique regular content that extends μ.*

It turns out that a regular content μ is always an exact function; i.e., for all compact sets A and B with $B \subset A$, $\mu(A) - \mu(B) = sup\{\mu(C): C \subset A \setminus B, C \in \mathscr{C}\}$.

9 PROPOSITION *Each regular content is an exact function.*

PROOF Suppose that A and B are compact sets and $B \subset A$. If C is a compact subset of $A \setminus B$ then $\mu(A) \geqq \mu(B \cup C) = \mu(B) + \mu(C)$. On the other hand, for $e > 0$ there is a compact set D so that $D^0 \supset B$ and $\mu(D) - \mu(B) < e$, whence, if $C = A \setminus D^0$, then $C \subset A \setminus B$ and

$\mu(A) \leqq \mu(D) + \mu(C) < \mu(B) + \mu(C) + e$. It follows that $\mu(A) = \mu(B) + sup\{\mu(C): C \subset A \setminus B$ and C compact$\}$. ■

A net $\{E_\alpha\}_{\alpha \in D}$ of sets is **decreasing** iff $E_\beta \subset E_\alpha$ whenever β follows α. A content μ is **hypercontinuous** iff $\mu(\bigcap_{\alpha \in D} E_\alpha) = lim_{\alpha \in D}\mu(E_\alpha)$ for every decreasing net $\{E_\alpha\}_\alpha$ in the family $\mathscr{C}$ of compact subsets of X. Since a content μ is monotonic, $lim_{\alpha \in D}\mu(E_\alpha) = inf_{\alpha \in D}\mu(E_\alpha)$ for a decreasing net $\{E_\alpha\}_\alpha$.

10 PROPOSITION *A content μ on $\mathscr{C}$ is regular iff it is exact, and this is the case iff it is hypercontinuous.*

PROOF We know a regular content is exact and we show that an exact content μ is regular. Suppose $B \in \mathscr{C}$ and $e > 0$, and let A be any compact neighborhood of B. Because μ is exact there is a compact subset C of $A \setminus B$ such that $\mu(A) < \mu(B) + \mu(C) + e$. Then every compact subset D of $A \setminus (B \cup C)$ has μ content less than e because $\mu(A) \geqq \mu(B \cup C \cup D) = \mu(B) + \mu(C) + \mu(D) > \mu(B) + (\mu(A) - \mu(B) - e) + \mu(D)$, whence $0 > -e + \mu(D)$. Let E be a compact neighborhood of C that is disjoint from B and let $F = A \setminus E^0$. Then F is a compact neighborhood of B, and if K is a compact subset of $F \setminus B$, then it is also a subset of $A \setminus (B \cup C)$ so $\mu(K) < e$. Taking the supremum of $\mu(K)$ for such K and using exactness, we find $\mu(F) - \mu(B) \leqq e$, so μ is regular.

We next show that if μ is regular, then it is hypercontinuous. Suppose $\{E_\alpha\}_{\alpha \in D}$ is a decreasing net of compact sets and $E = \bigcap_{\alpha \in D} E_\alpha$. For $e > 0$ choose a compact neighborhood F of E so that $\mu(F) < \mu(E) + e$. Since $\bigcap_{\alpha \in D} E_\alpha \subset F^0$, and each E_α is compact and F^0 is open, there is some finite subset $\{\alpha_1, \alpha_2, \ldots \alpha_n\}$ so $\bigcap_{i=1}^n E_{\alpha_i} \subset F^0$, and since D is directed, there is α so $E_\alpha \subset F^0$, whence $inf_{\alpha \in D}\mu(E_\alpha) \leqq \mu(F) < \mu(E) + e$. Thus μ is hypercontinuous.

Finally, suppose μ is hypercontinuous and $B \in \mathscr{C}$. Then the family D of compact neighborhoods α of B is directed by $\subset$, and if $E_\alpha = \alpha$ for each α, then $\{E_\alpha\}_{\alpha \in D}$ is decreasing and $\bigcap_{\alpha \in D} E_\alpha = B$. By hypercontinuity $lim_{\alpha \in D}\mu(E_\alpha) = \mu(\bigcap_{\alpha \in D} E_\alpha) = \mu(B)$, so there are compact neighborhoods of B with μ content near $\mu(B)$. Thus μ is regular. ■

SUPPLEMENT: G INVARIANT CONTENTS

We **suppose throughout that X is a locally compact Hausdorff space**, that **G is a group**, and that G **acts on X** in the following sense. For each a in G there is a homeomorphism (usually denoted $x \mapsto ax$ for x in X) of X onto X such that the map $x \mapsto ax$ followed by $x \mapsto bx$ is $x \mapsto (ba)x$; that is, the composition $(x \mapsto bx) \circ (x \mapsto ax)$ is $x \mapsto (ba)x$. Restated: If we let $\varphi(a)(x) = ax$, then φ is a homomorphism ($\varphi(ab) = \varphi(a) \circ \varphi(b)$) of G into the group of homeomorphisms of X onto itself. The situation is also described by saying X is a **left G space**. (If X is a **right**

G space and $\varphi(a)(x) = xa$, then $\varphi(ab) = \varphi(b) \circ \varphi(a)$; i.e., φ is an anti-homomorphism. If X is a left G space, then the definition $xa = a^{-1}x$ makes X a right G space.)

We also **assume** throughout that G acts **transitively** (for x and y in X there is a in G such that $ax = y$); and that the action of G is **semi-rigid** in the sense that if A and B are disjoint compact subsets of X and $x_0 \in X$, then there is a neighborhood V of x_0 such that no set of the form $\{aV = av: v \in V\}$ intersects both A and B. The group of rigid motions of $\mathbb{R}^n$ is the prototypical example of a semi-rigid transitive action.

A content μ for X is **G invariant** iff $\mu(aA) = \mu(A)$ for each a in G and for each compact subset A of X. We will show that there is a G invariant, regular content for X that is not identically zero.

Let us call a set of the form $aB = \{ax: x \in B\}$ a **G image**, or just an image of B, and for each subset E of G let $EB = \{ax: a \in E \text{ and } x \in B\}$. We begin the construction of a G invariant content by adopting a notation for the number of G images of a compact set B with $B^0 \neq \varnothing$ required to cover a compact set A. Let $[A|B]$ be the smallest number n such that there is a subset E of G with n members with $A \subset EB$. Notice that $[A|B][B|C] \geqq [A|C]$, for if $A \subset EB$ and $B \subset FC$ then $A \subset EFC$. Clearly $[aA|B] = [A|B]$ for each a in G.

We construct an approximation to a G invariant content from the function $(A, B) \mapsto [A|B]$ as follows. Let B be a fixed compact subset of X with non-void interior, and let x_0 be a fixed member of X. For each compact neighborhood V of x_0 and each member C of the class $\mathscr{C}$ of compact subsets of X, let $\lambda_V(C) = [C|V]/[B|V]$. Then λ_V has the following properties. It is non-negative, subadditive and monotone, and is G invariant in the sense that $\lambda_V(aC) = \lambda_V(C)$ for all a in G and C in $\mathscr{C}$. Moreover, $\lambda_V(\varnothing) = 0$ and $[C|B] \geqq \lambda_V(C) \geqq 1/[B|C]$ because $[C|B] \times [B|V] \geqq [C|V]$ and $[B|C][C|V] \geqq [B|V]$.

The function λ_V may fail to be additive, but it does have a sort of additive property: if no G image of V intersects both C and D, then $\lambda_V(C \cup D) = \lambda_V(C) + \lambda_V(D)$.

11 LEMMA *Let B be a compact subset of X with non-empty interior. Then there is a G invariant content λ on $\mathscr{C}$ such that $[C|B] \geqq \lambda(C) \geqq 1/[B|C]$ for all C in $\mathscr{C}$.*

PROOF For x_0 in X and a compact neighborhood V of x_0 let Z_V be the set of all monotone, G invariant, subadditive functions λ on $\mathscr{C}$ such that $[C|B] \geqq \lambda(C) \geqq 1/[B|C]$ for all C in $\mathscr{C}$, and such that λ is V additive in the sense that $\lambda(C \cup D) = \lambda(C) + \lambda(D)$ whenever no G image of V intersects both C and D. The set Z_V is not empty because the function λ_V constructed earlier is a member. Moreover, it is easy to check that Z_V is a closed subset of the product space $\times\{[0: [C|B]]: C \in \mathscr{C}\}$, this product is compact by Tychonoff's theorem, and so Z_V is compact. If $V \subset W$ then $Z_V \subset Z_W$, and therefore the family $\{Z_V$: *V a compact neighborhood of* $x_0\}$ has the finite intersection property. Consequently there is a member λ which belongs to Z_V for all V. That is, $\lambda(C \cup D) = \lambda(C) + \lambda(D)$ if there is some neighborhood V of x_0 such that no G image of V intersects both C and D. But the action of G is supposed to be semi-rigid by hypothesis, so λ is additive and the lemma is proved. ■

The G invariant content whose existence was just established is not identically zero because $\lambda(C) \geqq 1/[B|C]$, whence $\lambda(B) \geqq 1$. It may be that λ fails to be regular, but according to proposition 7, the regularization λ' given by $\lambda'(C) = inf\{\lambda(B)$: B is a compact neighborhood of $C\}$ is a regular content, and it is evidently G invariant. This establishes the following.

12 THEOREM *If the action of a group G by homeomorphisms on a locally compact Hausdorff space X is transitive and semi-rigid, then there is a regular G invariant content for X that is not identically zero.*

A **topological group** is a group G with a topology such that $x \mapsto x^{-1}$ is a continuous map of G onto itself and $(x, y) \mapsto xy$ is a continuous map of $G \times G$, with the product topology, into G. This is the same thing as requiring that $(x, y) \mapsto xy^{-1}$ be a continuous map of $G \times G$ into G. Notice that since the map $x \mapsto x^{-1}$ is its own inverse and is continuous, it is in fact a homeomorphism of G onto G.

If A and B are subsets of G, then $\boldsymbol{AB} = \{x$: *$x = yz$ for some y in A and some z in B*$\}$, and $\boldsymbol{A^{-1}} = \{x$: $x^{-1} \in A\}$. For each member b of G, $\boldsymbol{bA} = \{b\}A$ and $\boldsymbol{Ab} = A\{b\}$. The set bA is the **left translate** of A by b and Ab is the **right translate**. **Left translation** by b is the map $x \to bx$ of G, and **right translation** by b is $x \mapsto xb$. Left translation by b is continuous because it is the map $x \mapsto (b, x)$ followed by $(y, z) \mapsto yz$ and, since left translation by b^{-1} is the inverse, it is a homeomorphism. Consequently the left (or right) translate of a set which is compact, or open or closed, is of the same sort. If A is open, then A^{-1}, $BA = \bigcup_{b \in B} bA$ and $AB = \bigcup_{b \in B} Ab$ are open.

Each group G acts by left translation on itself; the assignment of a member a of G to the function $x \mapsto ax$ for x in G is such an action. Evidently the action is transitive (since $y = (yx^{-1})x$). We show that if G is a locally compact Hausdorff topological group, then the action by left translation is semi-rigid.

13 PROPOSITION *The action by left translation of a locally compact Hausdorff topological group G upon itelf is semi-rigid.*

Consequently there is a regular content μ for G, not identically zero, *which is left invariant in the sense that* $\mu(A) = \mu(aA)$ *for* A *compact and* a *in* G.

PROOF Suppose C and D are disjoint compact sets and that V is a neighborhood of the identity e such that for some x, $(xV) \cap C \neq \varnothing$ and $xV \cap D \neq \varnothing$; that is, $xv \in C$ and $xw \in D$ for some members v and w of V. Then $v^{-1}w \in C^{-1}D$ and consequently $(V^{-1}V) \cap (C^{-1}D) \neq \varnothing$. But $C^{-1}D$ is compact and $e \notin C^{-1}D$ because $C \cap D = \varnothing$. Consequently there is a neighborhood of (e, e) of the form $W^{-1} \times W$ whose image under the group map, $(y, z) \mapsto yz$, is disjoint from $C^{-1}D$. That is, $(W^{-1}W) \cap (C^{-1}D) = \varnothing$, and so no left translate of W intersects both C and D. ■

SUPPLEMENT: CARATHÉODORY PRE-MEASURES

Here is the classical Carathéodory construction for measures and of extending certain pre-measures. We assume throughout that **ν is a non-negative $\mathbb{R}^*$ valued function on the class $\mathscr{P}(X)$ of all subsets of X and that $\nu(\varnothing) = 0$**. Let us agree that a subset M of X **splits A additively** iff $\nu(A) = \nu(A \cap M) + \nu(A \setminus M)$. A set M is **Carathéodory ν measurable** iff M splits every member of $\mathscr{P}(X)$ additively. Let $\mathscr{M}$ be the family of Carathéodory measurable sets. Evidently $\varnothing \in \mathscr{M}$, $X \in \mathscr{M}$ and if $M \in \mathscr{M}$ then $X \setminus M \in \mathscr{M}$. We show that if M and N are members of $\mathscr{M}$ then so are $M \cap N$ and $M \cup N$ besides $X \setminus M$ (thus $\mathscr{M}$ is a **field** of sets), and if $M \cap N = \varnothing$, then $\nu(M \cup N) = \nu(M) + \nu(N)$ (that is, ν is **finitely additive** or just additive).

14 CARATHÉODORY LEMMA *The class $\mathscr{M}$ is a field of sets and $\nu | \mathscr{M}$ is finitely additive.*

PROOF We show that if M and N belong to $\mathscr{M}$ then $M \cap N \in \mathscr{M}$; it will follow that $\mathscr{M}$ is a field of sets. For each A in $\mathscr{P}(X)$, $\nu(A) = \nu(A \cap M) + \nu(A \setminus M)$ because A splits M additively, and $\nu(A \cap M) = \nu(A \cap M \cap N) + \nu((A \cap M) \setminus N)$ because N splits $A \cap M$ additively. But $\nu((A \cap M) \setminus N) + \nu(A \setminus M) = \nu(A \setminus (M \cap N))$ because M splits $A \setminus (M \cap N)$ additively, so $\nu(A) = \nu(A \cap M \cap N) + \nu(A \setminus (M \cap N))$ and hence $M \cap N \in \mathscr{M}$.

If M and N are disjoint members of $\mathscr{M}$, then $\nu(M \cup N) = \nu((M \cup N) \cap M) + \nu((M \cup N) \setminus M) = \nu(M + \nu(N)$. Thus ν is additive on the field $\mathscr{M}$ of sets. ■

We agree that ν is **countably subadditive** iff $\nu(A) \leqq \sum_k \nu(A_k)$ for all A and $\{A_n\}_n$ in the domain of ν such that $\{A_n\}_n$ covers A, and ν is **countably additive** iff $\nu(A) = \sum_k \nu(A_k)$ provided $\{A_n\}_n$ is a disjoint sequence with $\bigcup_n A_n = A$. If ν is countably sub-additive, then it is also finitely subadditive and monotonic. A family $\mathscr{A}$ of sets is called a **σ-field** iff $\mathscr{A}$ is a field of sets and $\mathscr{A}$ is closed under countable union.

15 THEOREM *If ν is countably subadditive non-negative $\mathbb{R}^*$ valued on $\mathscr{P}(X)$ and $\nu(\varnothing) = 0$, then the family $\mathscr{M}$ of Carathéodory ν measurable sets is a σ-field and ν is countably additive on $\mathscr{M}$. In fact $M \mapsto \nu(A \cap M)$, for M in $\mathscr{M}$, is countably additive for each $A \subset X$.*

PROOF If M and N are disjoint members of $\mathscr{M}$ and $A \subset X$, then $\nu(A \cap (M \cup N)) = \nu(A \cap (M \cup N) \cap M) + \nu((A \cap (M \cup N)) \setminus N) = \nu(A \cap M) + \nu(A \cap N)$ because $M \in \mathscr{M}$, so $M \mapsto \nu(A \cap M)$ is finitely additive on $\mathscr{M}$.

Suppose that $\{M_n\}_n$ is a disjoint sequence in $\mathscr{M}$ and $A \subset X$. Then for each n, $\nu(A) = \nu(A \setminus \bigcup_{k=1}^n M_k) + \nu(A \cap \bigcup_{k=1}^n M_k) \geqq \nu(A \setminus \bigcup_n M_n) + \sum_{k=1}^n \nu(A \cap M_k)$. Hence $\nu(A) \geqq \nu(A \setminus \bigcup_n M_n) + \sum_n \nu(A \cap M_n)$ so $\nu(A) \geqq \nu(A \setminus \bigcup_n M_n) + \nu(A \cap \bigcup_n M_n)$ because ν is countably sub-additive. But this last inequality is an equality because ν is subadditive, and we infer that $\bigcup_n M_n \in \mathscr{M}$ and (replace A by $A \cap \bigcup_n M_n$) that $M \mapsto \nu(A \cap M)$ is countably additive on $\mathscr{M}$. ■

The preceding theorem underlies an extension process whereby, under certain circumstances, one may extend a function $\mu: \mathscr{A} \to \mathbb{R}^+$ to a measure. Suppose that $\mathscr{A}$ is a family of subsets of X, μ on $\mathscr{A}$ is $\mathbb{R}^*$ valued non-negative, $\varnothing \in \mathscr{A}$, and $\mu(\varnothing) = 0$. The **outer measure μ^* induced by μ** is defined on $\mathscr{P}(X)$ by agreeing that $\mu^*(E) = \infty$ if no sequence in $\mathscr{A}$ covers E, and $\mu^*(E) = \inf\{\sum_n \mu(A_n)$: $\{A_n\}_n$ in $\mathscr{A}$ and $E \subset \bigcup_n A_n\}$ otherwise. Evidently μ^* is an extension of μ iff μ is countably sub-additive, and it is straightforward to verify that μ^* is itself countably subadditive (the "$\varepsilon/2^n$ argument"). Hence (taking $\nu = \mu^*$), if $\mathscr{M}$ is the family of Carathéodory measurable sets, μ^* is countably additive on $\mathscr{M}$. But an assumption must be made to ensure that $\mathscr{A} \subset \mathscr{M}$ if $\mu^* | \mathscr{M}$ is to be an extension of μ.

A function $\mu: \mathscr{A} \to \mathbb{R}^+$ is a **Carathéodory pre-measure** iff it is countably subadditive, $\mathscr{A}$ is non empty and closed under finite intersection and $\mu(A) = \mu(B) + \mu^*(A \setminus B)$ for all A and B in $\mathscr{A}$ with $B \subset A$.

16 EXTENSION THEOREM *Each Carathéodory pre-measure $\mu: \mathscr{A} \to \mathbb{R}^+$ extends to a countably additive $\mathbb{R}^*$ valued function μ^* on the σ-field $\mathscr{M}$ of Carathéodory μ^* measurable sets.*

PROOF We show that each member A of $\mathscr{A}$ is Carathéodory μ^* measurable. Suppose $B \subset X$ and $\{C_n\}_n$ is a sequence in $\mathscr{A}$ that covers B. Then $\mu(C_n) = \mu(C_n \cap A) + \mu^*(C_n \setminus A)$ for each n, so $\sum_n \mu(C_n) = \sum_n \mu(C_n \cap A) + \sum_n \mu^*(C_n \setminus A) \geqq \mu^*(B \cap A) + \mu^*(B \setminus A)$. Upon taking the infimum for all such sequences $\{C_n\}_n$, we obtain $\mu^*(B) \geqq \mu^*(B \cap A) + \mu^*(B \setminus A)$. The same inequality holds if no sequence in $\mathscr{A}$ covers B because in this case $\mu^*(B) = \infty$. The reverse inequality holds because μ^* is sub-additive, and it follows that $A \in \mathscr{M}$. ■

If μ is a Carathéodory pre-measure on $\mathscr{A}$ and $\mathscr{B} = \{B: B \in \mathscr{M}$ *and* $\mu^*(B) < \infty\}$, then $\mathscr{B}$ is a **δ-ring** of sets (i.e. $\mathscr{B}$ is closed under union, difference and countable intersection) and $\mu^*|\mathscr{B}$ is a non-negative, countably additive, real valued extension of μ. That is, $\mu^*|\mathscr{B}$ is a **measure** that is an extension of μ.

Note The Carathéodory condition, $\mu(A) - \mu(B) = \mu^*(A \setminus B)$ for all A and B in $\mathscr{A}$ with $B \subset A$ is, in a certain sense, the dual of the requirement for exactness: $\mu(A) - \mu(B) = sup\{\mu(C): C \subset A \setminus B\}$.

It is not difficult to see that a regular content (in the sense of the preceding supplements) is a Carathéodory pre-measure as well as exact. Consequently each regular content has an extension that is a measure. We shall also deduce this fact later, from exactness. We shall not use the Carathéodory results in what follows.

Chapter 2

PRE-MEASURE TO PRE-INTEGRAL

Each length function λ induces a rudimentary integration process as follows. If the function $\chi_{[a:b]}$ is 1 on the interval $[a:b]$ and 0 elsewhere, then its "integral" $I^\lambda(\chi_{[a:b]})$ with respect to λ should be $\lambda[a:b]$, and if $f = \sum_{i=1}^n c_i \chi_{[a_i:b_i]}$ then $I^\lambda(f)$ should be $\sum_{i=1}^n c_i \lambda[a_i:b_i]$. But is this assignment non-ambiguous? Stated in another way: does the function $\chi_{[a:b]} \mapsto \lambda[a:b]$ have a linear extension to the vector space of linear combinations of functions of the form $\chi_{[a:b]}$? It turns out that this is the case, and that it is a consequence of the fact that λ has an additive extension to a ring of sets containing the closed intervals, as we presently demonstrate.

A **ring of sets** is a non-empty family $\mathscr{A}$ of sets such that if A and B are members of $\mathscr{A}$ then $A \cup B$ and $A \setminus B$ also belong to $\mathscr{A}$. In other words, a non-empty family $\mathscr{A}$ of sets is a ring iff it is closed under difference and finite union.

The family $\mathscr{P}(X)$ of all subsets of a set X is a ring, as is the family of all finite subsets of X and the family of all countable subsets of X. Another example of a ring: the family of all finite unions of half-open intervals $(a:b]$, where a and b are real numbers and $(a:b] = \{x: a < x \leqq b\}$.

A ring $\mathscr{A}$ of sets is automatically closed under intersection because $A \cap B = A \setminus (A \setminus B)$, and it is also closed under symmetric difference because $A \triangle B = (A \setminus B) \cup (B \setminus A)$. Thus if $\mathscr{A}$ is a ring of subsets of X then $(\mathscr{A}, \triangle, \cap)$ is a ring in the algebraic sense; it is a subring of $\mathscr{P}(X)$ ($\triangle$ is the operation of ring addition; $\cap$ is multiplication). Conversely, a family $\mathscr{A}$ of sets which is closed under intersection and symmetric difference is closed under union and difference because $A \cup B = (A \triangle B) \triangle (A \cap B)$

and $A \setminus B = A \cap (A \,\triangle\, B)$. Hence a family of sets is a ring of sets iff $(\mathscr{A}, \triangle, \cap)$ is a ring in the algebraic sense. If in addition, $X = \bigcup \{A: A \in \mathscr{A}\} \in \mathscr{A}$ then $\mathscr{A}$ is a ring with unit X, or $\mathscr{A}$ is a **field of sets** for X, or just a **field of sets** or an **algebra of sets**.

If $\mathscr{A}$ is a family of sets then the family of all subsets of $\bigcup_{A \in \mathscr{A}} A$ is a ring that contains $\mathscr{A}$. The smallest ring that contains $\mathscr{A}$ is called the **ring generated by** $\mathscr{A}$; its members are just those sets that belong to every ring of sets that contains $\mathscr{A}$. Similarly the **lattice generated** by $\mathscr{A}$ is the smallest **lattice** (family closed under finite union and intersection) that contains $\mathscr{A}$.

It is not difficult to give a simple, explicit description of the ring $\mathscr{R}$ generated by a finite family $\{A_1, A_2, \ldots, A_n\}$. Let $X = \bigcup_{i=1}^n A_i$, let $A_i' = X \setminus A_i$ for each i, and for each subset M of $\{1, 2, \ldots n\}$ let $E_M = \bigcap_{j \in M} A_j \cap \bigcap_{j \notin M} A_j' = \bigcap_{j \in M} A_j \setminus \bigcup_{j \notin M} A_j$. Then E_M belongs to each ring of sets containing $\{A_1, A_2, \ldots, A_n\}$ and so $E_m \in \mathscr{R}$. If N is another subset of $\{1, 2, \ldots, n\}$ and $j \in M \setminus N$, then $E_m \subset A_j$ and $E_N \subset A_j'$ so E_M and E_N are disjoint, and consequently the family of all unions of sets of the form E_M is a ring $\mathscr{R}'$ with $\mathscr{R}' \subset \mathscr{R}$. But if $x \in A_i$ and $M = \{j: x \in A_j\}$, then $x \in E_M \subset A_i$, so A_i is the union of the sets E_M that it contains. Consequently $A_i \in \mathscr{R}'$, and it follows that $\mathscr{R}' = \mathscr{R}$. Thus $\mathscr{R}$ consists of unions of sets of the form E_M.

A non-empty set E_M is an **atom** of the ring $\mathscr{R}$, in the sense that $E_M \in \mathscr{R}$ and $\varnothing$ is the only member of $\mathscr{R}$ that is a proper subset of E_M. Thus $\mathscr{R}$ consists of all possible unions of atoms, i.e., $\mathscr{R}$ is **atomic**. Notice that each atom $E_M = \bigcap_{j \in M} A_j \setminus \bigcup_{j \notin M} A_j$ is the difference of members of the lattice $\mathscr{L}$ generated by $\{A_1, A_2, \ldots, A_n\}$. This establishes all except the last statement of the following.

1 LEMMA *If $\mathscr{L}$ is the lattice and $\mathscr{R}$ the ring generated by $\{A_1, A_2, \ldots, A_n\}$ then $\mathscr{R}$ is atomic, each atom is the difference of members of $\mathscr{L}$, and for each A in $\mathscr{L}$ there is B in $\mathscr{L}$ so $A \setminus B$ is an atom.*

PROOF Suppose that A is a non-empty member of $\mathscr{L}$. Choose a subset M of $\{1, 2, \ldots, n\}$ which is maximal with respect to the property that $\{A_j\}_{j \in M}$ fails to cover A and let $B = A \cap \bigcup_{j \in M} A_j$. If $k \notin M$ then $A_k \supset A \setminus \bigcup_{j \in M} A_j \neq \varnothing$ by maximality, and if $k \in M$ then $A_k' \supset A \setminus \bigcup_{j \in M} A_j$. Hence the atom $\bigcap_{k \notin M} A_k \cap \bigcap_{k \in M} A_k' \supset A \setminus B \neq \varnothing$, so $A \setminus B \in \mathscr{R}$ and is a non-empty subset of an atom of $\mathscr{R}$, and so must be identical with that atom. ■

We recall that an exact function μ is a real valued function on a lattice $\mathscr{A}$ of sets such that $\varnothing \in \mathscr{A}$, $\mu(\varnothing) = 0$, and $\mu(A) - \mu(B) = sup\{\mu(C): C \in \mathscr{A}$ and $C \subset A \setminus B\}$ for all A and B in $\mathscr{A}$ with $B \subset A$. For such sets A and B the number $\mu(A) - \mu(B)$ depends only on the difference set $A \setminus B$, and since $A \cup B \setminus B = A \setminus A \cap B$ for all A and B, $\mu(A \cup B) - \mu(B) = \mu(A) - \mu(A \cap B)$ for all A and B.

A function μ is **modular** iff it is a real valued function on a lattice $\mathscr{A}$ of sets, $\varnothing \in \mathscr{A}$, $\mu(\varnothing) = 0$ and $\mu(A) + \mu(B) = \mu(A \cup B) + \mu(A \cap B)$ for all A and B in $\mathscr{A}$. Each exact function is modular. A real (or real*) valued function μ on a family $\mathscr{A}$ is **finitely additive** iff $\mu(A \cup B) = \mu(A) + \mu(B)$ for all disjoint members A and B of $\mathscr{A}$ with $A \cup B$ in $\mathscr{A}$. The following proposition asserts that each modular function on $\mathscr{A}$, and in particular each exact function, has a finitely additive real valued extension to the ring generated by $\mathscr{A}$. (We owe the proof to H. v. Weizsäcker.)

2 THEOREM *The ring $\mathscr{R}$ generated by a lattice $\mathscr{A}$ of sets consists of unions of finitely many disjoint sets of the form $A \setminus B$ with A and B in $\mathscr{A}$.*

Each modular function μ on $\mathscr{A}$ has a unique finitely additive extension $\mu^{\sim}$ to $\mathscr{R}$.

PROOF The family $\mathscr{B} = \{B: B$ *belongs to the ring generated by some finite subfamily* $\{A_1, A_2, \ldots, A_n\}$ *of* $\mathscr{A}\}$ is itself a ring containing $\mathscr{A}$, and consequently $\mathscr{B} \supset \mathscr{R}$. The first assertion of the proposition then follows from lemma 1, and the proof of the second assertion reduces to the case where $\mathscr{A}$ is finite.

Let us define $\mu'(A \setminus B)$ to be $\mu(A) - \mu(B)$ for A and B in $\mathscr{A}$ with $B \subset A$. This definition is not ambiguous, for the following reasons. Suppose A, B, C and D belong to $\mathscr{A}$, $B \subset A$, $D \subset C$ and $A \setminus B = C \setminus D$. Then $A = (A \cap C) \cup B$ so, since μ is modular, $\mu(A) = \mu(A \cap C) + \mu(B) - \mu(A \cap C \cap B)$ and since $B \cap C = B \cap D$, we have $\mu(A) - \mu(B) = \mu(A \cap C) - \mu(B \cap D)$ which by symmetry is $\mu(C) - \mu(D)$. We note that any additive extension of μ must agree with μ' on differences, so an additive extension of μ to $\mathscr{R}$, if there is one, is unique.

Each atom of $\mathscr{R}$ is the difference of two members of $\mathscr{A}$, and we define $\mu^{\sim}$ of an arbitrary member A of $\mathscr{R}$ to be $\sum\{\mu'(T)$: $T \subset A$ *and* T *is an atom of* $\mathscr{R}\}$. Clearly, $\mu^{\sim}$ is additive on $\mathscr{R}$ and we show by induction on the number of atoms contained in A that $\mu^{\sim}(A) = \mu(A)$ for A in $\mathscr{A}$. If A is a non-empty member of $\mathscr{A}$, then there is, by the lemma, a member B of $\mathscr{A}$ such that $B \subset A$ and $A \setminus B$ is an atom, whence $\mu^{\sim}(A) - \mu^{\sim}(B) = \mu^{\sim}(A \setminus B) = \mu'(A \setminus B) = \mu(A) - \mu(B)$, so $\mu^{\sim}(A) - \mu^{\sim}(B) = \mu(A) - \mu(B)$. The inductive hypotheses implies that $\mu^{\sim}(B) = \mu(B)$ and so $\mu^{\sim}(A) = \mu(A)$. ■

A real* valued function μ on a family $\mathscr{A}$ of sets is **countably additive** iff $\mu(\sum_n A_n) = \sum_n \mu(A_n)$ for all disjoint sequences $\{A_n\}_n$ in $\mathscr{A}$ with $\bigcup_n A_n$ in $\mathscr{A}$. If $\mathscr{A}$ is a ring of sets then countable additivity can be viewed as finite additivity plus a continuity condition, as follows. If $\{A_n\}_n$ is a disjoint sequence in $\mathscr{A}$, then $\sum_n \mu(A_n) = lim_n \sum_{k=1}^n \mu(A_k)$, and since μ is finitely additive, this is $lim_n \mu(\bigcup_{k=1}^n A_k)$. Consequently μ is countably additive iff $lim_n \mu(\bigcup_{k=1}^n A_k) = \mu(\bigcup_n A_n)$ for every disjoint sequence in $\mathscr{A}$ with $\bigcup_n A_n$ in $\mathscr{A}$. The sequence $\{\bigcup_{k=1}^n A_k\}_n$ of partial unions of $\{A_n\}_n$ is an increasing sequence in $\mathscr{A}$, and every increasing sequence $\{B_n\}_n$ in $\mathscr{A}$ is the sequence of partial unions of the disjoint sequence B_1, $B_2 \setminus B_1$, $B_3 \setminus B_2$, ... in $\mathscr{A}$. It follows that μ is countably additive iff it is **continuous from below**, in the sense that $lim_n \mu(B_n) = \mu(\bigcup_n B_n)$ for each increasing sequence $\{B_n\}_n$ in $\mathscr{A}$ with $\bigcup_n B_n$ in $\mathscr{A}$. (A sequence $\{B_n\}_n$ is **increasing** iff $B_n \subset B_{n+1}$ for all n, and it is **decreasing** iff $B_n \supset B_{n+1}$ for all n.)

There are other characterizations of countable additivity. We agree that a real* valued function μ on $\mathscr{A}$ is **continuous from above at** A iff $A \in \mathscr{A}$ and $\mu(A) = lim_n \mu(A_n)$ for each decreasing sequence $\{A_n\}_n$ in $\mathscr{A}$ with $A = \bigcap_n A_n$ and $\mu(A_1) < \infty$, that μ is **continuous from above** iff it is continuous from above at each member A of $\mathscr{A}$, and that μ is **continuous at** $\varnothing$ iff it is continuous from above at $\varnothing$. We notice that the counting function γ, which assigns to a set A of integers the number $\gamma(A)$ of members of A, is continuous from above, although γ is ∞ on each of the sets $\{k: k > n\}$ but is zero on their intersection.

Lastly, μ is **subadditive** iff $\mu(C) \leqq \mu(A) + \mu(B)$ for A, B and C in $\mathscr{A}$ with $C \subset A \cup B$, and μ is **countably sub-additive** iff $\mu(C) \leqq \sum_n \mu(A_n)$ provided $C \in \mathscr{A}$, $\{A_n\}_n$ is a sequence in $\mathscr{A}$, and $C \subset \bigcup_n A_n$.

We observe that if μ is non-negative and finitely additive on a ring $\mathscr{A}$ then it is **monotonic** in the sense that $\mu(B) \geqq \mu(A)$ if $B \supset A$, because $\mu(B) = \mu(A) + \mu(B \setminus A)$.

3 PROPOSITION *If μ is a real* valued, finitely additive, non-negative function on a ring $\mathscr{A}$ of set, then μ is countably additive iff it is continuous from below, and this is the case iff it is countably subadditive.*

If μ is countably additive it is continuous from above and if μ is finite valued and continuous at $\varnothing$ then it is countably additive.

PROOF We have already seen that μ is countably additive iff it is continuous from below. Suppose μ is countably additive, $\{A_n\}_n$ is a sequence in $\mathscr{A}$, $C \subset \bigcup_n A_n$ and $C \in \mathscr{A}$. Let $B_n = \bigcup_{k=1}^n C \cap A_k$ for each n. Then $\{B_n\}_n$ is increasing and $C = \bigcup_n B_n$, so $B_1, B_2 \setminus B_1, B_3 \setminus B_2, \ldots$ are disjoint and $\mu(C) = \mu(B_1) + \sum_n \mu(B_{n+1} \setminus B_n) \leqq \mu(A_1) + \sum_n \mu(A_{n+1})$ because $B_1 \subset A_1$ and $B_{n+1} \setminus B_n \subset A_{n+1}$ for each n. Consequently μ is countably subadditive.

If $\{A_n\}_n$ is a disjoint sequence in $\mathscr{A}$ and $\bigcup_n A_n \in \mathscr{A}$ then $\sum_n \mu(A_n) = lim_n \sum_{k=1}^n \mu(A_k) = lim_n \mu(\bigcup_{k=1}^n A_k) \leqq \mu(\bigcup_n A_n)$. If μ is countably subadditive then $\sum_n \mu(A_n) \geqq \mu(\bigcup_n A_n)$, whence $\sum_n \mu(A_n) = \mu(\bigcup_n A_n)$ and μ is countably additive.

Suppose $\{A_n\}_n$ is a decreasing sequence in $\mathscr{A}$, $A = \prod_n A_n \in \mathscr{A}$, and $\mu(A_1) < \infty$. Then $\{A_1 \setminus A_n\}_n$ is increasing, the union is $A_1 \setminus A$, and $\mu(A_1 \setminus A_n) = \mu(A_1) - \mu(A_n)$ for each n because $\mu(A_1 \setminus A_n) + \mu(A_n) =$

$\mu(A_1)$ and $\mu(A_n) \leqq \mu(A_1) < \infty$. Therefore, if μ is countably additive, and hence continuous from below, then $\mu(A_1) - \mu(A) = lim_n(\mu(A_1) - \mu(A_n))$ so $\mu(A) = lim_n \mu(A_n)$. Thus μ is continuous from above.

Finally, suppose that μ is finite valued and continuous from above at $\varnothing$ and that $\{A_n\}_n$ is a disjoint sequence in $\mathscr{A}$ with $A = \bigcup_n A_n \in \mathscr{A}$. Then $\{A \setminus \bigcup_{k=1}^n A_k\}_n$ is a decreasing sequence with void intersection so $lim_n \mu(A \setminus \bigcup_{k=1}^n A_k) = 0$. It follows that $\mu(A) = lim_n \mu(\bigcup_{k=1}^n A_k) = lim_n \sum_{k=1}^n \mu(A_k)$, so μ is countably additive. ■

We have notice that each exact function μ on a lattice $\mathscr{A}$ of sets is modular, and so by theorem 2 it has a unique finitely additive extension $\mu^\sim$ to the ring $\mathscr{R}$ generated by $\mathscr{A}$. If A and B belong to $\mathscr{A}$ and $B \subset A$, then $\mu^\sim(A \setminus B) = \mu(A) - \mu(B) = sup\{\mu(C): C \in \mathscr{A} \text{ and } C \subset A \setminus B\}$, and since every member of $\mathscr{R}$ is the union of finitely many disjoint sets of the form $A \setminus B$, $\mu^\sim(R) = sup\{\mu(C): C \in \mathscr{A} \text{ and } C \subset R\}$ for all R in $\mathscr{R}$. We use this fact to show that $\mu^\sim$ is continuous at $\varnothing$ if μ is.

4 PROPOSITION *The unique finitely additive extension $\mu^\sim$ on $\mathscr{R}$ of an exact function μ on $\mathscr{A}$ is given by $\mu^\sim(R) = \sup\{\mu(A): A \in \mathscr{A}$ and $A \subset R\}$.*

If μ is continuous at $\varnothing$ then so is $\mu^\sim$.

PROOF The first statement has already been established.

Suppose that $\{R_n\}_n$ is a decreasing sequence in $\mathscr{R}$ with $\bigcap_n R_n = \varnothing$ and that $\{e_n\}_n$ is a sequence of positive numbers with $\sum_n e_n$ small. For each n there is A_n in $\mathscr{A}$ with $A_n \subset R_n$ and $\mu^\sim(R_n) - \mu(A_n) < e_n$. We show inductively that $\mu^\sim(R_p) - \mu(\bigcap_{i=1}^p A_i) < e_1 + e_2 + \cdots + e_p$ for each p, and since $\{\bigcap_{i=1}^p A_i\}_p$ is a decreasing sequence in $\mathscr{A}$ with void intersection, it will follow that $\mu^\sim(R_p)$ is arbitrarily small for p sufficiently large.

The proposition is clear for $p = 1$. The inductive step: by modularity, $\mu(A_{p+1}) + \mu(\bigcap_{i=1}^p A_i) = \mu(A_{p+1} \cap \bigcap_{i=1}^p A_i) + \mu(A_{p+1} \cup \bigcap_{i=1}^p A_i) \leqq \mu(\bigcap_{i=1}^{p+1} A_i) + \mu^\sim(R_p) < \mu(\bigcap_{i=1}^{p+1} A_i) + \mu(\bigcap_{i=1}^p A_i) + e_1 + e_2 + \cdots + e_p$ by the inductive hypothesis. So $\mu(A_{p+1}) < \mu(\bigcap_{i=1}^{p+1} A_i) + e_1 + e_2 + \cdots + e_p$ and $\mu^\sim(R_{p+1}) < \mu(\bigcap_{i=1}^{p+1} A_i) + e_1 + e_2 + \cdots + e_p + e_{p+1}$. ■

If μ is exact on a lattice $\mathscr{A}$ of compact sets then μ is automatically continuous at $\varnothing$, for if $\{A_n\}_n$ is decreasing in $\mathscr{A}$ with $\bigcap_n A_n = \varnothing$, then $A_n = \varnothing$ for n large. Hence:

5 COROLLARY *Each exact function μ on a lattice of compact sets has a countably additive, non-negative real valued extension $\mu^\sim$ to the ring $\mathscr{R}$ generated by $\mathscr{A}$, given by $\mu^\sim(R) = \sup\{\mu(A): A \in \mathscr{A}$ and $A \subset R\}$ for R in $\mathscr{R}$.*

A function μ on a family $\mathscr{A}$ of sets is called a **pre-measure** provided $\mathscr{A}$ is closed under finite intersection and μ has a countably additive, non-negative real valued extension to the ring $\mathscr{R}$ generated by $\mathscr{A}$. Thus each exact function that is continuous at $\varnothing$ is a pre-measure. The preceding corollary, together with theorem 1.6, shows that every length function is a pre-measure. (The term "pre-measure" is used in anticipation of the theorem that every pre-measure has an extension that is a measure.)

A **characteristic function** is a function that assumes no value other than 0 or 1. If $A \subset X$ then the **characteristic function of** A (on X), denoted χ_A, is defined to be 1 at points of A and 0 at points of $X \setminus A$.

A real valued function f is **simple** iff it has finite range or, equivalently, iff f is a finite linear combination of characteristic functions. If $\mathscr{A}$ is a family of subsets of X, then a function f on X is $\mathscr{A}$ **simple** iff f is a finite linear combination of characteristic functions of members of $\mathscr{A}$. We denote by $L^{\mathscr{A}}$ the vector space of $\mathscr{A}$ simple functions. If $\mathscr{A}$ is a lattice of sets then $L^{\mathscr{A}}$ is identical with the class $L^{\mathscr{R}}$ of $\mathscr{R}$ simple functions, where $\mathscr{R}$ is the ring generated by $\mathscr{A}$, because $\mathscr{R}$ consists of unions of finitely many disjoint sets of the form $A \setminus B$ with A and B in $\mathscr{A}$. Further, if $\mathscr{B}$ is a family closed under intersection and $\mathscr{A}$ is the lattice generated by $\mathscr{B}$ then $L^{\mathscr{B}} = L^{\mathscr{A}} = L^{\mathscr{R}}$. We omit the straightforward proof of this fact.

If $\mathscr{A}$ is a ring of sets then each $\mathscr{A}$ simple function is linear combination of characteristic functions of *disjoint* members of $\mathscr{A}$ (e.g., $a\chi_A + b\chi_B = a\chi_{A\setminus B} + (a+b)\chi_{A\cap B} + b\chi_{B\setminus A}$). Further, if f and g belong to $L^{\mathscr{A}}$, then we may suppose $f = \sum_{k=1}^m a_k \chi_{C_k}$ and $g = \sum_{k=1}^m b_k \chi_{C_k}$ for some disjoint finite sequence $\{C_1, C_2, \ldots, C_m\}$ in $\mathscr{A}$. It follows that $f \vee g$ and $f \wedge g$, where $(f \vee g)(x) = max\{f(x), g(x)\}$ and $(f \wedge g)(x) = min\{f(x), g(x)\}$, belong to $L^{\mathscr{A}}$, so $L^{\mathscr{A}}$ is a vector function lattice. This is a lattice with **truncation**, in the sense that if $f \in L^{\mathscr{A}}$ then $1 \wedge f \in L^{\mathscr{A}}$. Thus $L^{\mathscr{A}}$ is a vector function lattice with truncation. (We think of $L^{\mathscr{A}}$ as a linearization of the ring $\mathscr{A}$.)

Suppose that μ is a modular function on a lattice $\mathscr{A}$. We will construct a linear functional I^μ on $L^{\mathscr{A}}$ such that $I^\mu(\chi_A) = \mu(A)$ for all A in $\mathscr{A}$. It turns out that if μ is monotonic then I^μ is **positive**, in the sense that $I^\mu(f) \geqq 0$ if $f \geqq 0$. If μ is a pre-measure, $I^\mu(f)$ will turn out to be the integral of f with respect to any measure that is an extension of μ.

6 PROPOSITION *If μ is a modular function on a lattice $\mathscr{A}$ of sets, then there is a unique linear functional I^μ on $L^{\mathscr{A}}$ such that $I^\mu(\chi_A) = \mu(A)$ for A in $\mathscr{A}$, and if μ is monotonic then I^μ is positive.*

PROOF Suppose for the moment that μ is finitely additive on a ring $\mathscr{A}$ of sets. We show that if $A_1, A_2, \ldots, A_n$ are members of $\mathscr{A}$ and $\sum_{i=1}^n a_i \chi_{A_i} = 0$, then $\sum_{i=1}^n a_i \mu(A_i) = 0$, whence the definition

$I^\mu(\sum_{i=1}^n a_i\chi_{A_i}) = \sum_{i=1}^n a_i\mu(A_i)$ is not ambiguous. Choose a finite disjoint subfamily $\mathscr{B}$ of the ring $\mathscr{A}$ so that $A_i = \bigcup\{B: B \in \mathscr{B}$ *and* $B \subset A_i\}$ for each i. Then $\mu(A_i) = \sum\{\mu(B): B \in \mathscr{B}$ *and* $B \subset A_i\}$ and $\sum_{i=1}^n a_i\mu(A_i) = \sum_{B\in\mathscr{B}}\mu(B)\ (\sum\{a_i: B \subset A_i\})$. But for B in $\mathscr{B}$, $\sum\{a_i: B \subset A_i\} = 0$ because $\sum_{i=1}^n a_i\chi_{A_i} = 0$. Consequently $\sum_{i=1}^n a_i\mu(A_i) = 0$.

If μ is modular on a lattice $\mathscr{A}$ then μ extends, by theorem 2, to a finitely additive function $\mu^\sim$ on the ring $\mathscr{R}$ generated by $\mathscr{A}$, and $I^{\mu^\sim}(\chi_A) = \mu(A)$ for A in $\mathscr{A}$. Each member of $\mathscr{R}$ is the union of finitely many disjoint sets of the form $A \setminus B$ with $\varnothing \subset B \subset A$ and A and B in $\mathscr{A}$, and if μ is monotonic then $\mu^\sim$ is nonnegative and $I^{\mu^\sim}$ is positive. Uniqueness of $I^{\mu^\sim}$ follows from the fact that $\{\chi_A: A \in \mathscr{A}\}$ spans $L^{\mathscr{A}}$. ■

Here are some useful consequences of positivity of I^μ. If f and g belong to L and $f \geqq g$, then $I^\mu(f - g) \geqq 0$ so $I^\mu(f) \geqq I^\mu(g)$. If $0 \leqq g \leqq a\chi_A$ with A in $\mathscr{A}$ then $I^\mu(g) \leqq a\mu(A)$. A set S is a **support** for f iff f is zero outside S. If S is a support for f, $f \in L^{\mathscr{A}}$ and $S \in \mathscr{A}$ then $I^\mu(f) \leqq \mu(S)\,max\,f$.

A **pre-integral**, or a **Daniell–Stone pre-integral** is a positive linear functional I on a vector function lattice L with truncation, such that $lim_n I(f_n) = 0$ for every decreasing sequence $\{f_n\}_n$ in L that converges pointwise to zero. This last condition can be stated alternatively: I is a **countably additive linear functional**, in the sense that $I(\sum_n g_n) = \sum_n I(g_n)$ for all sequences $\{g_n\}_n$ of non-negative members of L with pointwise sum $\sum_n g_n$ belonging to L. (The proof is the usual partial sum and differencing trick.) Thus a pre-integral is a positive, countably additive linear functional on a vector function lattice with truncation.

7 THEOREM (PRE-MEASURE TO PRE-INTEGRAL) *If μ is a pre-measure on $\mathscr{A}$, then I^μ is a pre-integral on $L^{\mathscr{A}}$.*

PROOF Suppose μ is a pre-measure on $\mathscr{A}$ and that $\{f_n\}_n$ is a decreasing sequence in $L^{\mathscr{A}}$ with $lim_n f_n = 0$. We may suppose that $\mathscr{A}$ is a ring of sets. Let S in $\mathscr{A}$ be a support for f_1; let b be an upper bound for f_1, and for $e > 0$ let $A_n = \{x: f_n(x) > e\}$. Then $\{A_n\}_n$ is decreasing and $\bigcap_n A_n = \varnothing$, so $\mu(A_n)$ is small for n large because μ is continuous at $\varnothing$. Hence for n large, $f_n < e$ save on a set A with $\mu(A) < e$, $f_n \leqq b\chi_A + e\chi_S$ and $I^\mu(f_n) < b\mu(A) + e\mu(S) < (b + \mu(S))e$. Consequently $lim_n I^\mu(f_n) = 0$. ■

There are pre-integrals that are not induced by pre-measures. Let $\boldsymbol{C[a\!:\!b]}$ be the set of all continuous real valued functions f on $[a\!:\!b]$. Then the Riemann integral $R(f) = \int_a^b f(t)\,dt$, for f in $C[a\!:\!b]$, is a positive linear functional on a vector function lattice with truncation. We will show that it is a pre-integral.

There is a variant of this example. Let $\boldsymbol{C_c}(\mathbb{R})$ be the class of continuous real valued functions on $\mathbb{R}$ that have **compact supports** (i.e., for each f there is a compact set K such that f is identically zero on $\mathbb{R}\setminus K$). Then $C_c(\mathbb{R})$ is a vector function lattice with truncation and the Riemann integral, restricted to $C_c(\mathbb{R})$, is a pre-integral. We prove this after establishing a preliminary result. Recall that a real valued function f is **upper semi-continuous**, or **u.s.c.** iff $\{x: f(x) \geqq c\}$ is closed for each c in $\mathbb{R}$, and f is **lower semi-continuous** iff $\{x: f(x) \leqq c\}$ is closed for each c.

8 DINI'S THEOREM *If a decreasing sequence $\{f_n\}_n$ of u.s.c. functions on $[a:b]$, or on $\mathbb{R}$ with compact supports, converges pointwise to zero, then it converges to zero uniformly.*

PROOF For $e > 0$ let $A_n = \{x: f_n(x) \geqq e\}$. Then $\{A_n\}_n$ is a decreasing sequence of compact sets and $\bigcap_n A_n = \varnothing$ because $lim_n f_n(x) = 0$ for all x. Consequently there is n such that $A_n = \varnothing$. It follows that $\{f_n\}_n$ converges uniformly to zero. ■

9 PROPOSITION *Each positive linear functional, and in particular the Riemann integral, on either $C[a:b]$ or $C_c(\mathbb{R})$, is a pre-integral.*

PROOF We prove only that a positive linear functional on $C_c(\mathbb{R})$ is a pre-integral. Suppose $\{f_n\}_n$ is a decreasing sequence in $C_c(\mathbb{R})$, and let $\| f_n \| = sup_{x \in \mathbb{R}} |f(x)|$ for each n. By Dini's theorem, $lim_n \| f_n \| = 0$.

Choose M so $|f_1(x)| = 0$ if $|x| \geqq M$ and choose a non-negative member h of $C_c(\mathbb{R})$ that is 1 on $[-M:M]$, whence $f_n \leqq \| f_n \| h$ for all n. Then $0 \leqq I(f_n) \leqq \| f_n \| I(h)$ because I is positive, and hence $lim_n I(f_n) = 0$. ■

In the next chapter we will extend each pre-integral to an integral. An **integral** is a pre-integral, say J on M, with the additional **Beppo Levi property**: If $\{f_n\}_n$ is an increasing sequence in M that converges pointwise to a real valued function f, and if $sup_n J(f_n) < \infty$, then $f \in M$ and $J(f) = lim_n J(f_n)$. In chapter 4 we show that each integral induces a measure so that each pre-measure induces a pre-integral, then an integral, and finally a measure.

SUPPLEMENT: VOLUME λ_n; THE ITERATED INTEGRAL

A closed interval in $\mathbb{R}$ is a set of the form $A = [a:b]$; a **closed interval in $\mathbb{R}^n$**, or an ***n*-interval**, is the Cartesian product $\times_{i=1}^n A_i$ of n closed intervals $\{A_i\}_{i=1}^n$ in $\mathbb{R}$, and the class of n-intervals is denoted by $\mathscr{I}_n$. The ***n*-dimensional volume** or just the **volume** is defined on $\mathscr{I}_n$ by setting $\lambda_n(\times_{i=1}^n A_i)$ equal to the product $\prod_{i=1}^n \ell(A_i)$, where ℓ is the usual length. Thus $\lambda_1 = \ell$, and $\lambda_{n+1}(\mathsf{X}_{i=1}^{n+1} [a_i : b_i]) = \prod_{i=1}^{n+1} (b_i - a_i) = \lambda_n(\mathsf{X}_{i=1}^{n} [a_i : b_i])\ell(a_{n+1} : b_{n+1})$. We shall not distinguish between $\mathsf{X}_{i=1}^{n+1} A_i$ and $(\mathsf{X}_{i=1}^{n} A_i) \times A_{n+1}$.

Let L^n $(=L^{\mathscr{J}_n})$ be the class of $\mathscr{J}_n$ simple functions. Because $\mathscr{J}_n$ is closed under intersection, L^n is identical with the class of $\mathscr{R}$ simple functions, where $\mathscr{R}$ is the ring generated by $\mathscr{J}_n$ (we observe that $\chi_{A \cup B} = \chi_A + \chi_B - \chi_{A \cap B}$). Consequently L^n is a vector function lattice with truncation.

Suppose $f = \sum_{i=1}^{k} a_i \chi_{A_i}$ for some $a_1, a_2, \ldots, a_k$ in $\mathbb{R}$ and $A_1, A_2, \ldots, A_k$ in $\mathscr{J}_n$. We define $I^n(f)$ to be $\sum_{i=1}^{k} a_i \lambda_n(A_i)$. The first assertion of the following proposition implies that this definition is not ambiguous, and that I^n is a positive linear functional on L^n.

10 PROPOSITION *If $a_i \in \mathbb{R}$ and $A_i \in \mathscr{J}_n$ for $i = 1, 2, \ldots, k$, and if $\sum_{i=1}^{k} a_i \chi_{A_i} \geqq 0$, then $\sum_{i=1}^{k} a_i \lambda_n(A_i) \geqq 0$.*
The function I^n is a pre-integral on L^n.

PROOF The proposition is true for $n = 1$, and we argue by induction. Each n-interval A_i is the cartesian product $B_i \times C_i$ with B_i in $\mathscr{J}_{n-1}$ and C_i in $\mathscr{J}_1$, and if $(x, y) \in \mathbb{R}^{n-1} \times \mathbb{R}^1$, then $\chi_{A_i}(x, y) = \chi_{B_i}(x)\chi_{C_i}(y)$. For each x in $\mathbb{R}^{n-1}$ the function $\sum_{i=1}^{k} a_i \chi_{B_i}(x)\chi_{C_i}$ is a non-negative $\mathscr{J}_1$ simple function on $\mathbb{R}$, and hence $\sum_{i=1}^{k} a_i \lambda_1(C_i)\chi_{B_i}(x) \geqq 0$. Consequently, by the induction hypothesis, $0 \leqq \sum_{i=1}^{k} a_i \lambda_1(C_i)\lambda_{n-1}(B_i) = \sum_{i=1}^{k} a_i \lambda_n(A_i)$.

Evidently I^n is a positive linear functional on L^n, and we show by induction that it is a pre-integral. Suppose that $\{s_k\}_k$ is a descending sequence in L^n that converges pointwise to zero. Then for each k, $s_k = \sum_{i=1}^{p_k} a_{ik}\chi_{A_{ik}}$ with $A_{ik} = B_{ik} \times C_{ik}$, with B_{ik} in $\mathscr{J}_{n-1}$ and C_{ik} in $\mathscr{J}_1$. For each x in $\mathbb{R}^{n-1}$, $\{\sum_{i=1}^{p_k} a_{ik}\chi_{B_{ik}}(x)\chi_{C_{ik}}\}_k$ is a decreasing sequence of $\mathscr{J}_1$ simple functions on $\mathbb{R}$ that converges pointwise to zero, whence $\{\sum_{i=1}^{p_k} a_{ik}\lambda_1(C_{ik})\chi_{B_{ik}}(x)\}_k$ is decreasing and converges to zero for each x, and hence, by the induction assumption $\{\sum_{i=1}^{p_k} a_{ik}\lambda_1(C_{ik})\lambda_{n-1}(B_{ik})\}_k = \{I^n(s_k)\}_k$ converges to zero. ∎

The Riemann integral restricted to $C_c(\mathbb{R})$ is a pre-integral R according to corollary 9. There is a similar pre-integral on $C_c(\mathbb{R}^n)$, the class of continuous real valued functions on $\mathbb{R}^n$ that have compact supports. It is defined recursively as follows. For f in $C_c(\mathbb{R}^{n+1})$ and y in $\mathbb{R}$, let $f_y(x_1, x_2, \ldots, x_n) = f(x_1, x_2, \ldots, x_n, y)$. We let $R_1 = R$. If $f \in C_c(\mathbb{R}^{n+1})$ then $f_y \in C_c(\mathbb{R}^n)$ for all y, $y \mapsto R_n(f_y)$ belongs to $C_c(\mathbb{R})$, and $R_{n+1}(f)$ is defined to be $R_1(y \mapsto R_n(f_y))$. The functional R_n is the nth **iterated Riemann integral**. Evidently it is a positive linear functional. We show that it is a pre-integral. (This fact has already been established for $n = 1$ and the generalization to locally compact Hausdorff spaces is theorem 12 in the next supplement.)

11 PROPOSITION *Every positive linear functional I on $C_c(\mathbb{R}^n)$, and in particular the iterated Riemann integral, is a pre-integral.*

PROOF Suppose a decreasing sequence $\{f_k\}_k$ in $C_c(\mathbb{R}^n)$ converges to zero pointwise. Then the sequence converges to zero uniformly by the argument in Dini's theorem (theorem 8), so if $\| f_k \| = sup_{x \in \mathbb{R}^n} |f_k(x)|$ then $\{ \| f_k \| \}_k$ converges to zero.

Choose M so that if $x \in \mathbb{R}^n$ and $|x_i| \geqq M$ for $i = 1, 2, \ldots, n$, then $f_1(x)$, and hence $f_q(x)$ for all q, is zero. If P is the Cartesian product of n copies of $[-M:M]$, then $f_q \leq \| f_q \| \chi_P$ and evidently $\chi_P \leqq h = \prod_{i=1}^n ((1 + M - |x_i|) \vee 0)$. Then $h \in C_c(\mathbb{R}^n)$, $f_q \leqq \| f_q \| h$, so $0 \leqq I(f_q) \leqq \| f_q \| I(h)$ because I is positive, and hence $lim_q I(f_q) = 0$. ■

SUPPLEMENT: PRE-INTEGRALS ON $C_c(X)$ AND $C_0(X)$

We assume, for the rest of this section, that X is a **locally compact Hausdorff space.**

Let $C_c(X)$ be the family of all real valued continuous functions on X that have compact supports; and let $C_0(X)$ be the family of all real valued continuous functions that **vanish at** ∞ in the sense that for $e > 0$, there is a compact subset K of X such that $|f(x)| < e$ for each member x of $X \setminus K$.

12 PROPOSITION *If I is a positive linear functional on either $C_c(X)$ or $C_0(X)$, then I is a pre-integral.*

PROOF Suppose $\{f_n\}_n$ is a decreasing sequence in $C_c(X)$ that converges to 0 pointwise, and that K is a compact support for f_1. Then it is easy to see that $\{f_n\}_n$ converges uniformly to zero on K—indeed, the argument for theorem 8, as written, establishes this. Choose a non-negative number g of $C_c(X)$ such that $g \geqq \chi_K$. (Urysohn's lemma, applied to a compact neighborhood V of K, shows that there is such a function g.) Then $f_n \leqq (sup_x f_n(x))\chi_K \leqq (sup_x f_n(x))g$ and hence $I(f_n) \leqq (sup_x f_n(x))I(g)$. Hence $lim_n I(f_n) = 0$, and consequently I is a pre-integral on $C_c(X)$.

Suppose that I is a positive linear functional on $C_0(X)$ and that $Y = X \cup \{\infty\}$ is the one point compactification of X (the open neighborhoods of ∞ are complements of compact subsets of X). For each f in $C(Y)$ let $J(f) = f(\infty) + I(f|X)$. Then J is a positive linear functional on $C(Y) = C_c(Y)$ and is consequently a pre-integral. Finally, if $\{f_n\}_n$ is a sequence in $C_0(X)$ that converges pointwise to zero, $g_n(y) = f_n(y)$ for y in X and $g_n(\infty) = 0$, then $0 = lim_n J(g_n) = lim_n I(f_n)$. Thus I is a pre-integral on $C_0(X)$. ■

The validity of the preceding proposition requires that the domain of I be large enough. A positive linear functional I on a vector sublattice of $C_c(X)$ or $C_0(X)$ need not be a pre-integral—for example, let L be the family of all continuous functions on $[0:1]$ which vanish at 0 and which have a right derivative there, and for f in L let $I(f)$ be the value of the right derivative at 0.

There is an extension of the preceding proposition that applies to bounded linear functionals on $C_0(X)$. The supremum **norm** of a member f of $C_0(X)$ is defined by $\| f \|_X = sup_{x \in X} |f(x)|$, and a linear functional F on $C_0(X)$ is **bounded** relative to the sup norm iff $\|F\| = sup\{|F(f)|: \| f \|_X \leqq 1\} < \infty$. One sees without difficulty that $|F(f)| \leqq \|F\| \, \| f \|_X$.

The norm $\| \ \|_X$ induces a metric $(f, g) \mapsto \| f - g \|_X$ for $C_0(X)$, and the bounded linear functionals are just those that are continuous relative to the metric topology. The space $C_0(X)$ is complete since convergence relative to the metric is just uniform convergence.

The following argument shows that each pre-integral I on $C_0(X)$ is bounded, whence so is the difference of two pre-integrals on $C_0(X)$. Suppose $\| f_n \|_X \leqq 1$ and $I(f_n) \geqq 2^n$ for each n. Then, since $I(|f_n| - f_n) \geqq 0$, $I(|f_n|) \geqq 2^n$. The sequence $\{\sum_{n=1}^N 2^{-n} |f_n|\}_N$ is a Cauchy sequence and converges to some $f \geqq \sum_{n=1}^N 2^{-n} |f_n|$, whence $I(f) \geqq \sum_{n=1}^N 2^{-n} I(|f_n|) \geqq N$ for all N, and this is a contradiction.

13 PROPOSITION *A linear functional F on $C_0(X)$ is bounded iff it is the difference of two pre-integrals.*

PROOF It is only necessary to show that a bounded linear functional F is the difference of two positive linear functional, and this will follow (see chapter 0) provided we show that $F^+(f) = sup\{F(u): u \in C_0(X) \text{ and } 0 \leqq u \leqq f)\} < \infty$ for all non-negative members f of $C_0(X)$. If $0 \leqq u \leqq f$, then $|F(u)| \leqq \|F\| \, \|u\|_X \leqq \|F\| \, \| f \|_X$ whence $F^+(f) \leqq \|F\| \, \| f \|_X < \infty$. ■

Chapter 3

PRE-INTEGRAL TO INTEGRAL

This section is devoted to the construction of an integral from a pre-integral, and to a few consequences. Among these consequences are norm completeness, Fatou's lemma, the monotone convergence theorem and the dominated convergence theorem for an arbitrary integral.

We recall that a pre-integral is a positive linear functional I on a vector function lattice L with truncation such that $lim_n I(f_n) = 0$ for every decreasing sequence $\{f_n\}_n$ in L that converges pointwise to zero. This last condition is equivalent to requiring that $lim_n I(f_n) = I(f)$ for every increasing (or decreasing) sequence $\{f_n\}_n$ in L that converges pointwise to a member f of L. An alternative statement: I is countably additive in the sense that $\sum_n I(f_n) = I(f)$ for every sequence $\{f_n\}_n$ of non-negative members of L with pointwise sum $f = \sum_n f_n$ belonging to L.

An **integral** is a positive linear functional I on a vector function lattice L with truncation that has the property (the Beppo Levi property): if $\{f_n\}_n$ is an increasing sequence in L, $sup_n I(f_n) < \infty$, and $f(x) = sup_n f_n(x) < \infty$ for all x, then $f \in L$ and $I(f) = lim_n I(f_n)$. An alternative statement: if $\{f_n\}_n$ is a sequence of non-negative members of L with $\sum_n I(f_n) < \infty$ and $f(x) = \sum_n f_n(x) < \infty$ for all x, then $f \in L$ and $I(f) = \sum_n I(f_n)$. Notice that the domain of an integral is closed under pointwise convergence of decreasing sequences of non-negative members. Each integral is evidently also a pre-integral.

If I is any positive linear functional on L and f and g are members of L with $f \leqq g$ then $I(g - f) \geqq 0$ so $I(f) \leqq I(g)$. Consequently $|I(f)| \leqq I(|f|)$ because $-|f| \leqq f \leqq |f|$. The **norm** (or L_1 **norm**) of a member f

of L, $\| f \|_1$, or just $\| \ \|$, is defined to be $I(|f|)$. Thus $|I(f)| \leqq \| f \|_1$. (It would be more precise to call $\| \ \|_1$ the "norm induced by I", and label it "$\| \ \|_I$".)

The L_1 norm is unfortunately not always a norm. A **norm** for a vector space E is a real valued non-negative function $\| \ \|$ on E such that

(i) $\| f + g \| \leqq \| f \| + \| g \|$ for all f and g in E,
(ii) $\| t f \| = |t| \, \| f \|$ for all f in E and all scalars t, and
(iii) for each f in E, if $\| f \| = 0$ then $f = 0$.

The L_1 norm has properties (i) and (ii) but, in general, may fail to satisfy (iii). It should properly be called a **semi-norm** or a **pseudo-norm**, but we follow time honored usage in calling it a norm.

If $\| \ \|$ is a semi-norm for a vector space E then $\| f - g \|$ is the **norm distance** from f to g, and the function $(f, g) \mapsto \| f - g \|$ is the **norm semi-metric**. A sequence $\{f_n\}_n$ is **fundamental** or **Cauchy** iff $lim_{m,n} \| f_m - f_n \| = 0$, and E is **complete** iff each Cauchy sequence in E converges to some member of E. This is the case iff each Cauchy sequence in E has a subsequence which converges to a member of E.

The space L of a pre-integral I on L may fail to be complete, and the integral induced by I is to be a completion. That is, we enlarge L to a space L^1, and extend I on L to I^1 on L^1 so that I^1 is a pre-integral *and* L^1 is complete relative to its natural norm. The obvious approach is to adjoin to L the pointwise limits of Cauchy sequences in L, but unfortunately a Cauchy sequence may fail to converge at any point. For example, the characteristic functions of the intervals, $[0:1]$, $[0:1/2]$, $[1/2:1]$, $[0:1/3]$, $[1/3:2/3]$, $[2/3:1]$, $[0:1/4]$, ... converge at no point of $[0:1]$. But this sequence of characteristic functions is Cauchy relative to the norm $f \mapsto I(|f|)$, if I is the Riemann integral on the class of piecewise continuous functions on $[0:1]$. Thus we cannot hope to complete L by adjoining pointwise limits of arbitrary Cauchy sequences. But a variant of this idea works.

A sequence $\{g_n\}_n$ in a semi-normed space E is **swiftly convergent** iff $\sum_n \| g_{n+1} - g_n \| < \infty$. A swiftly convergent sequence $\{g_n\}_n$ is automatically a Cauchy sequence because $\| g_{n+p+1} - g_n \| = \| \sum_{k=n}^{n+p} (g_{k+1} - g_k) \| \leqq \sum_{k=n}^{\infty} \| g_{k+1} - g_k \|$, and this sum is small for n large. Each Cauchy sequence has a swiftly convergent subsequence, and it follows that E is complete iff each swiftly convergent sequence in E converges relative to the norm distance to a member of E.

There is a natural one to one correspondence between swiftly convergent sequences and sequences with summable norms. If a sequence $\{f_n\}_n$ in E has **summable norms** in the sense that $\sum_n \| f_n \| < \infty$, then the sequence $\{\sum_{k=1}^{n} f_k\}_n$ of partial sums is swiftly convergent. On the other hand, if $\{g_n\}_n$ is swiftly convergent then it is the sequence of partial sums of the sequence $g_1, g_2 - g_1, \ldots, g_{n+1} - g_n, \ldots$, which has summable norms. We agree that a sequence $\{f_n\}_n$ is **norm summable to**

f iff the sums of $\{f_n\}_n$ over finite subsets of $\mathbb{N}$ converge to f (explicitly, for $e > 0$ there is a finite subset F of $\mathbb{N}$ such that $\|\sum_{n \in G} f_n - f\| < e$ for all finite G with $F \subset G \subset \mathbb{N}$). If $\{f_n\}_n$ is norm summable to f, then $lim_n \|f - \sum_{k=1}^n f_k\| = 0$.

Statements about swiftly convergent sequences can always be translated into statements about sequences with summable norms, and vice versa. The following proposition is an example. (A methodological note: swift convergence is convenient for order theoretic arguments and summing sequences is convenient for arguments involving linearity.)

1 PROPOSITION *A semi-normed space E is complete iff each swiftly convergent sequence in E converges to a member of E, or iff each sequence in E with summable norms is norm summable to a member of E.*

We would like to extend a pre-integral I on L by adjoining to L the pointwise sums of sequences in L that have summable norms and then extending I to such sums. Unfortunately, such a sequence $\{f_n\}_n$ may not be summable at each point—the set $\{x: \sum_n |f_n(x)| = \infty\}$ may be non-void—but we cope with this exceptional set in statesmanlike fashion. We ignore it.

A subset E of X is **null**, or ***I* null**, iff there is a sequence $\{f_n\}_n$ in L with summable norms such that $\sum_n |f_n(x)| = \infty$ for each x in E (i.e., $E \subset \{x: \sum_n |f_n(x)| = \infty\}$). Thus, if $\sum_n \|f_n\| < \infty$, then $\{f_n(x)\}_n$ is (absolutely) summable except for members x of the I null set $\{x: \sum_n |f_n(x)| = \infty\}$, and each swiftly convergent sequence $\{g_n\}_n$ in L converges pointwise outside of the null set $\{x: \sum_n |g_{n+1}(x) - g_n(x)| = \infty\}$.

Each subset of an I null set is evidently I null. If $e > 0$ then $\{x: \sum_n |f_n(x)| = \infty\} = \{x: \sum_n |ef_n(x)| = \infty\}$, so a set E is I null iff there is $\{g_n\}_n$ in L with $\sum_n |g_n(x)| = \infty$ for each x in E and with $\sum_n \|g_n\| < e$. If, for each n, E_n is a null set, then there is a sequence $\{f_{n,k}\}_k$ in L with $\sum_k \|f_{n,k}\| < 2^{-n}$ and $\sum_k |f_{n,k}(x)| = \infty$ for all x in E_n, whence $\{f_{n,k}\}_{n,k}$ is a double sequence with summable norms and $\sum_{n,k} |f_{n,k}(x)| = \infty$ on $\bigcup_n E_n$; therefore the countable union of null sets is null. The same sort of reasoning shows that E is null iff for $e > 0$ there is $\{f_n\}_n$ in L with $\sum_n |f_n| \geqq \chi_E$ and $\sum_n \|f_n\| < e$. Here is yet another description of null sets: E is null iff there is an increasing sequence $\{h_n\}_n$ of non-negative members of L with $\sup_n \|h_n\| < \infty$ and $sup_n\, h_n(x) = \infty$ for x in E.

We emphasize the inconsequential nature of null sets by agreeing that a proposition about x holds **almost everywhere** or **for almost every *x*** (**a.e.**, for **a.e. *x***, ***I* a.e.**, for ***I* a.e. *x***) iff the set of points x for which the proposition fails is I null.

We extend the pre-integral I on L by adjoining all pointwise sums of sequences $\{f_n\}_n$ in L with summable norms. More precisely: we adjoin to L real valued functions f such that $f(x) = \sum_n f_n(x)$ for

I almost every x, and we define the integral of f to be $\sum_n I(f_n)$. The next lemma implies that this definition is not ambiguous.

2 FUNDAMENTAL LEMMA *If $\{f_n\}_n$ is a sequence in L, $\sum_n \|f_n\| < \infty$ and $\sum_n f_n(x) \geqq 0$ for I a.e. x, then $\infty > \sum_n I(f_n) \geqq 0$.*

If $\{g_n\}_n$ is a swiftly convergent sequence in L and $lim_n\, g_n(x) \geqq 0$ for I a.e.x then $\infty > lim_n\, I(g_n) \geqq 0$.

PROOF We first observe that if $\{u_n\}_n$ is a sequence of non-negative members of L and $\sum_n u_n \geqq v \in L$ then $\sum_n I(u_n) \geqq I(v)$ because $\sum_{k=1}^n u_k \geqq (\sum_{k=1}^n u_k) \wedge v$ so $\sum_{k=1}^n I(u_k) \geqq I((\sum_{k=1}^n u_k) \wedge v)$ and hence, taking limits, $\sum_{k=1}^\infty I(u_k) \geqq I(lim_n((\sum_{k=1}^n u_k) \wedge v)) = I(v)$. If $\sum_n u_n \geqq \sum_n v_n$, where $\{v_n\}_n$ is a sequence of non-negative members of L, then $\sum_n I(u_n) \geqq \sum_{k=1}^N I(v_k)$ for each N, and hence $\sum_n I(u_n) \geqq \sum_n I(v_n)$.

Suppose $\{f_n\}_n$ is a sequence in L with $\sum_n \|f\|_n < \infty$ and $\sum_n f_n(x) \geqq 0$ except for x in a null set E. We may choose a sequence $\{h_n\}_n$ of non-negative members of L with $\sum_n \|h_n\| < \infty$ such that $\sum_n h_n(x) = \infty$ for all x in E. If $f_n^+ = f_n \vee 0$ and $-f_n^- = f_n \wedge 0$, then $f_n = f_n^+ - f_n^-$ and $\|f_n\| \geqq max\{\|f_n^+\|, \|f_n^-\|\}$ so $\{f_n^+\}$ and $\{f_n^-\}_n$ have summable norms. For a fixed positive number t let $u_n = f_n^+ + th_n$ and $v_n = f_n^-$. Then $\sum_n u_n(x) \geqq \sum_n v_n(x)$ for all x and the preceding paragraph implies that $\sum_n I(u_n) = \sum_n I(f_n^+) + t\sum_n I(h_n) \geqq \sum_n I(v_n) = \sum_n I(f_n^-)$. Since this is the case for all positive t, it must be that $\sum_n I(f_n^+) \geqq \sum_n I(f_n^-)$ so $\sum_n I(f_n) \geqq 0$. ■

The **Daniell extension I^1 on L^1** of I on L is defined as follows. A real valued function f belongs to L^1, or is **I^1 integrable**, iff there is a sequence $\{f_n\}_n$ in L with summable norms such that $f(x) = \sum_n f_n(x)$ for I almost every x, and in this case $\boldsymbol{I^1(f)} = \sum_n I(f_n)$. If $\{g_n\}_n$ is also a sequence in L with summable norms and $f(x) = \sum_n g_n(x)$ I a.e.x, then $\sum_n (f_n - g_n) = 0 = \sum_n (g_n - f_n)$ I a.e., hence $\sum_n (I(f_n) - I(g_n)) = 0$ according to the fundamental lemma and $\sum_n I(f_n) = \sum_n I(g_n)$. Thus the definition of I^1 is not ambiguous. It follows from the definition that L^1 is a vector space and I^1 is a positive linear functional on that vector space.

The members f of L^1 can also be described as pointwise limits I a.e. of swiftly convergent sequences $\{f_n\}_n$ in L, and $I^1(f) = lim_n\, I(f_n)$. If $\{f_n\}_n$ and $\{g_n\}_n$ are swiftly convergent sequences in L, then so are $\{f_n \wedge g_n\}_n$, $\{f_n \vee g_n\}$ and $\{f_n \wedge 1\}_n$. It follows that L^1 is a vector function lattice with truncation. Hence I^1 on L^1 will be shown to be an integral if it has the Beppo Levi property: if $\{f_n\}_n$ is an increasing sequence in L^1, $f(x) = sup_n\, f_n(x)$ for all x, and $sup_n\, I^1(f_n) < \infty$, then $f \in L^1$ and $I^1(f) = lim_n\, I^1(f_n)$. We establish this fact after recording for reference a single lemma.

The norm $\|f\|_1$ of a member f of L^1 is $I^1(|f|)$—$|f| \in L^1$ because

L^1 is a vector lattice. Evidently $\| \;\|_1$ is an extension of the norm $\| \;\|$ for L. If $\{h_n\}_n$ is a sequence in L with summable norms, then $|\sum_n h_n| \leqq \sum_n |h_n|$ so $\|\sum_n h_n\|_1 = I^1(|\sum_n h_n|) \leqq (\sum_n I|h_n|) = \sum_n \|h_n\|$. Hence:

3 LEMMA *If I is a pre-integral on L, $\{f_n\}_n$ is a sequence in L with summable norms and $f(x) = \sum_n f_n(x)$ for I a.e. x, then $\|f - \sum_{k=1}^n f_k\| \leqq \sum_{k=n+1}^\infty \|f_k\|$.*

Consequently, if $f \in L^1$ and $e > 0$, then there is a sequence $\{g_n\}_n$ with summable norms in L, such that $f = \sum_n g_n$ I a.e., $\|f - g_1\|_1 < e$ and $\sum_n \|g_n\| < \|f\|_1 + e$.

Suppose that $\{f_n\}_n$ is a sequence of non-negative members of a Daniell extension L^1 with summable norms and that $f(x) = \sum_n f_n(x) < \infty$ for all x. Then there is, for each n, a sequence $\{f_{n,k}\}_k$ in L such that $\sum_k \|f_{n,k}\| < \|f_n\|_1 + 2^{-n}$ and $f_n = \sum_k f_{n,k}$ except on an I null set E_n. Then $\sum_{n,k} \|f_{n,k}\| < \infty$ and $f = \sum_n f_n = \sum_{n,k} f_{n,k}$ except on the I null set $E \cup \bigcup_n E_n$, where $E = \{x: \sum_{n,k} |f_{n,k}(x)| = \infty\}$. Consequently $f \in L^1$ and $I^1(f) = \sum_n I^1(f_n)$. Thus:

4 THEOREM (PRE-INTEGRAL TO INTEGRAL) *The Daniell extension of a pre-integral is an integral.*

A particular consequence of lemma 3 is that for f in L^1 and $e > 0$ there is g in L with $\|f - g\|_1 < e$; that is, L is **dense in** L^1. It is also the case that L^1 is complete, so that L^1 is a **completion** of L. In outline: if $\{f_n\}_n$ is a Cauchy sequence in L^1, there is $\{g_n\}_n$ in L so that $lim_n \|f_n - g_n\|_1 = 0$, $\{g_n\}_n$ is Cauchy and so has a swiftly convergent subsequence, this subsequence converges to a member f of L^1, hence so does $\{g_n\}_n$ and so does $\{f_n\}_n$.

We will presently show that the domain of *every* integral is complete. We use the Daniell extension process to deduce properties of an arbitrary integral.

5 LEMMA ON NULL SETS *If J is an integral on M, $g \in M$, $\{f_n\}_n$ is a sequence in M with summable norms and $E = \{x: \sum_n |f_n(x)| = \infty\}$, then $\chi_E \in M$, $g\chi_E \in M$ and $J(\chi_E) = J(g\chi_E) = 0$.*

PROOF We may assume without loss of generality that $f_n \geqq 0$ for each n, so the sequence $\{s_n\}_n$ of partial sums is increasing, $sup_n J(s_n) < \infty$ and $E = \{x: lim_n s_n(x) = \infty\}$. For each k, the sequence $\{(2^{-k}s_n) \wedge 1\}$ is increasing and $J((2^{-k}s_n) \wedge 1) \leqq 2^{-k}J(s_n)$. Consequently, since J is an integral, $lim_n((2^{-k}s_n) \wedge 1)$ belongs to M and J of the limit is bounded by $2^{-k}\, sup_n J(s_n)$. The sequence $\{lim_n(2^{-k}s_n) \wedge 1\}_k$ is decreasing, and it follows that $lim_k\, lim_n((2^{-k}s_n) \wedge 1)$ belongs to M and J of this double limit is zero. But $lim_k \, lim_n ((2^{-k} s_n(x)) \wedge 1)$ is 1 if $lim_n \, s_n(x) = \infty$ and 0 otherwise; that is, the double limit is χ_E, so $\chi_E \in M$ and $J(\chi_E) = 0$.

An even simpler argument serves to establish the assertion about g. We assume $g \geqq 0$. For each k, $J(g \wedge (2^k \chi_E)) \leqq J(2^k \chi_E) = 0$, and taking the limit on k yields $g\chi_E \in M$ and $J(g\chi_E) = 0$. ■

If $g \in M$, $g \geqq 0$ J a.e. and $J(g) = 0$, then the set $A = \{x : g(x) \neq 0\}$ and all its subsets are J null because the sequence $|g|, |g|, \ldots$ has summable norms and its pointwise sum is ∞ on A. In particular if $\chi_E \in M$ and $J(\chi_E) = 0$, the subsets of E are J null. The preceding lemma shows that all J null sets are of this form.

6 Theorem on Norm Completeness *If J is an integral on M, then each swiftly convergent sequence in M is dominated J a.e. by a member of M and converges J a.e. and in norm to a member of M.*

In particular M is norm complete.

PROOF Suppose J is an integral on M, that $\{f_n\}_n$ is a sequence in M with $\sum_n \|f_n\| < \infty$, $E = \{x: \sum_n |f_n(x)| = \infty\}$ and $f = \sum_n (1 - \chi_E) f_n$. Then f belongs to the Daniell extension M^1 of M and by lemma 3 the partial sums of $\{f_n\}_n$ converge J a.e. and in norm to f. We assert that when J is an integral on M, f belongs to M. This is the case because: if $f_k^+ = f_k \vee 0$, then by the preceding lemma, $(1 - \chi_E) f_k^+ \in M$ for each k, $\{\sum_{k=1}^n (1 - \chi_E) f_k^+\}_n$ is increasing, $J(\sum_{k=1}^n (1 - \chi_E) f_k^+ = J(\sum_{k=1}^n f_k^+) \leqq \sum_k J(|f_k|) < \infty$, whence $\sum_n (1 - \chi_E) f_n^+ \in M$, $\sum_n (1 - \chi_E) f_n^- \in M$ similarly, and so $f \in M$.

It follows that a swiftly convergent sequence $\{f_n\}_n$ in M converges J a.e. and in norm to a member of M. Since each Cauchy sequence in M has a swiftly convergent subsequence, M is norm complete. Moreover, the sequence $\{|f_{n+1} - f_n|\}_n$ has summable norms, so if $h(x) = |f_1(x)| + \sum_n |f_{n+1}(x) - f_n(x)|$ for x outside the set E where the sum on the right is infinite, and $h(x) = 0$ for x in E, then $h \in M$ and $\{f_n\}_n$ is dominated by h a.e. ■

The preceding theorem does *not* imply that every integral J on M is identical with its Daniell extension J^1 on M^1. A swiftly convergent sequence $\{f_n\}_n$ in M converges J a.e. and in norm to some member g of M, but it also converges J a.e. and in norm to $g + h$ for every real valued function h with J null support, and such a function h may fail to belong to M. In fact, the Daniell extension M^1 of an integral J on M is precisely $\{u$: *u is a real valued function agreeing J a.e. with some member g of M*$\} = \{g + h$: *$g \in M$, and h has J null support*$\}$

The integral J on M is said to be **null complete** iff each real valued function h with J null support belongs to M, or iff $M = M^1$. If an integral J on M is an extension of a pre-integral I on L, then J^1 on M^1

is an extension of I^1 on L^1, and if M is null complete so that $M = M^1$, then J on M is itself an extension of I^1 on L^1. Thus every null complete integral that extends a pre-integral I on L also extends I^1 on L^1. Assuming for the moment that a Daniell extension is always null complete (it is), we conclude that the Daniell extension of a pre-integral is its smallest null complete integral extension. Because of this, the Daniell extension of an *integral* I on L is called the **usual** or the **minimal null completion** of I on L; it is often denoted $\boldsymbol{I^\vee}$ on $\boldsymbol{L^\vee}$ as well as I^1 on L^1.

7 PROPOSITION *The Daniell extension of a pre-integral is its smallest null complete integral extension.*

In particular, every null complete integral that extends an integral I on L also extends its null completion $I^\vee$ on $L^\vee$.

PROOF We need only show that the Daniell extension of a pre-integral I on L is null complete. If h has I^1 null support E, then there is a sequence $\{f_n\}_n$ in L^1 with summable norms such that $E \subset \{x: \sum_n |f_n(x)| = \infty\}$. By lemma 3, for each n there is a sequence $\{f_{n,k}\}_k$ in L such that $\sum_n \|f_{n,k}\| < 2^{-n} + \|f_n\|_1$ and $f_n(x) = \sum_k f_{n,k}(x)$ outside an I null set E_n. Then $\{f_{n,k}\}_{n,k}$ has summable norms and $E \subset \{x: \sum_{n,k} |f_{n,k}(x)| = \infty\} \cup \bigcup_n E_n$. Hence E is I null, so $h \in L^1$ and the proposition is proved. ■

A sequence $\{f_n\}_n$ is **increasing $\boldsymbol{J}$ a.e.** iff J is an integral and for each n, $f_{n+1}(x) \geqq f_n(x)$ for J a.e. x. A J a.e. increasing sequence in M such that $sup_n J(f_n) < \infty$ is evidently swiftly convergent. The theorem on norm completeness therefore has the following corollary.

8 MONOTONE CONVERGENCE THEOREM *If J is an integral on M, $\{f_n\}_n$ is a sequence that is increasing J a.e. and* $\sup_n J(f_n) < \infty$, *then $\{f_n\}_n$ converges J a.e. and in norm to a member f of* M, *and* $J(f) = \lim_n J(f_n)$.

There is a useful consequence of the preceding result. Recall that a set V with partial ordering $\geqq$ is **order complete (conditionally complete, Dedekind complete)** iff each non-empty subset W which has an upper bound in V has a supremum in V (that is, a member w of V that is an upper bound for W and is less than or equal to every other upper bound).

Suppose J is an integral on M. Let us agree that $\boldsymbol{f \geqq_J g}$ iff f and g belong to M and $f \geqq g$ J almost everywhere. We show that M is $\geqq_J$ order complete. In fact, if J is bounded from above on a non-empty subset W of M that is closed under $\vee$ (i.e., if g and h are in W, then so is $g \vee h$), then there is an increasing sequence $\{f_n\}_n$ in W such that

$sup_n J(f_n) = sup\{J(f): f \in W\} < \infty$, theorem 8 shows that $\{f_n\}_n$ converges J a.e. to a member f of M, and it is straightforward to verify that f is *a* $\geqq_J$ supremum of W.

9 Corollary *If J is an integral on M, then a non-empty subset W of M that is closed under $\vee$ has a $\geqq_J$ supremum, provided* $\sup\{J(g): g \in W\} < \infty$.

In particular, M with the ordering $\geqq_J$, is order complete.

Thus the domain of an integral is both norm complete and order complete; it may or may not be null complete.

We will establish another convergence theorem for integrals after an important preliminary lemma. (A stronger form of the lemma is given in chapter 7.)

10 Fatou's Lemma *Suppose J is an integral on M* and $\{f_n\}_n$ *is a sequence of non-negative members of M such that* $\liminf_n J(f_n) < \infty$ *and* $\liminf_n f_n(x) < \infty$ *for all x.*

Then $\liminf_n f_n \in M$ *and* $J(\liminf_n f_n) \leqq \liminf_n J(f_n)$.

PROOF For each n, the sequence $f_n, f_n \wedge f_{n+1}, f_n \wedge f_{n+1} \wedge f_{n+2}, \ldots$ is decreasing and so its pointwise limit, $inf_{k \geqq n} f_k$, belongs to M because M is the domain of an integral. Since $inf_{k \geqq n} f_k \leqq f_n$, $J(inf_{k \geqq n} f_k) \leqq J(f_n)$ and so $lim_n J(inf_{k \geqq n} f_k) \leqq lim\,inf_n J(f_n) < \infty$. Consequently, the B. Levi property of integrals shows $lim\,inf_n f_n = lim_n inf_{k \geqq n} f_k \in M$ and $J(lim\,inf_n f_n) = lim_n J(inf_{k \geqq n} f_k)$, which is at most $lim\,inf_n J(f_n)$. ■

11 Dominated Convergence Theorem *Suppose that J is an integral on M, $g \in M$, $\{f_n\}_n$ is a sequence in M that converges pointwise to a function f, and $|f_n(x)| \leqq g(x)$ for all x.*

Then $f \in M$, $J(f) = \lim_n J(f_n)$ *and $\{f_n\}_n$ converges to f in norm.*

PROOF Fatou's lemma applied to the sequence $\{f_n + g\}_n$ shows that $lim\,inf_n(f_n + g)$ belongs to M and that $J(lim\,inf_n(f_n + g)) \leqq lim\,inf_n J(f_n + g)$. Consequently $f = lim_n f_n$ belongs to M *and* $J(f) \leqq lim\,inf_n J(f_n)$. The same lemma applied to the sequence $\{-f_n + g\}_n$ shows that $J(-f) \leqq lim\,inf_n J(-f_n) = -lim\,sup_n J(f_n)$, whence $J(f) = lim_n J(f_n)$.

This result, applied to the sequence $\{|f_n - f|\}_n$, which is dominated by $2g$, shows that $lim_n \|f_n - f\|_1 = 0$. ■

There is a slightly stronger form of the preceding, obtained by judiciously sprinkling "a.e.'s" almost everywhere. It reads: *If a sequence in M converges pointwise a.e. to f and is dominated a.e. by a member of M, then it converges a.e. and in norm to a member of M, and this member* *agrees a.e. with f*. This form can be easily obtained from theorem 11 by using lemma 5 (on null sets).

Both the monotone convergence and the dominated convergence theorems deduce "limit of the integral = integral of the limit" results from pointwise convergence plus an additional assumption. *Some* additional assumption is necessary, as the following simple examples show.

The first example is called the "**moving bump**". Let ℓ be length and let I^ℓ be the induced preintegral on the space $L^{\mathscr{I}}$ of $\mathscr{I}$ simple functions. If $f_n = \chi_{[n:n+1]}$ for each n, then $lim_n f_n = 0$ but $lim_n I^\ell(f_n) = 1 \neq 0 = I^\ell(lim_n f_n)$.

If all the members of the sequence $\{f_n\}_n$ vanish outside a fixed interval $[a:b]$, then pointwise convergence of $\{f_n\}_n$ to a member f of $L^{\mathscr{I}}$ still fails to imply convergence of $\{I^\ell(f_n)\}_n$ to $I^\ell(f)$, as the "**growing steeple**" example shows. Let $f_n = n\chi_{(0:1/n)}$ for each n. Then $lim_n f_n = 0$ but $lim_n I^\ell(f_n) = 1$. A further note: both the moving bump and growing steeple examples can be modified to get sequences of differentiable functions, with the same sort of behavior relative to the integral induced by I^ℓ.

It can happen that the Daniell extensions of different pre-integrals are identical—indeed the prototype of an integral occurs naturally as the Daniell extension of two quite different pre-integrals. We recall that the Riemann integral R on the space $C_c(\mathbb{R})$ of continuous real valued functions on $\mathbb{R}$ with compact support is a pre-integral according to proposition 2.9, and theorem 2.7 implies that the length function ℓ induces a pre-integral I^ℓ on the class $L^{\mathscr{I}}$ of linear combinations of finitely many characteristic function $\chi_{[a:b]}$.

12 PROPOSITION *The Riemann integral R on $C_c(\mathbb{R})$ and the pre-integral I^ℓ induced by length have the same Daniell extension.*

Because the Daniell extension of a pre-integral is its smallest null complete integral extension (proposition 7), it is sufficient to show that $(I^\ell)^1$ is an extension of the Riemann integral R on $C_c(\mathbb{R})$ and that R^1 is an extension of I^ℓ. We leave this proof to the reader.

Finally, the **Lebesgue integral I^1** on $L^1(\mathbb{R})$ is defined to be the Daniell extension of the pre-integral I^ℓ induced by length. The **Lebesgue integral** for an **interval** $[a:b]$, is defined by: $f \in L^1[a:b]$ iff the extension g of f which is zero on $\mathbb{R}\backslash[a:b]$ belongs to $L^1(\mathbb{R})$, and in this case $\int_a^b f(t)\,dt = I^1(g)$. It is, in fact, the case that this integral is also the Daniell extension of the Riemann integral on $C[a:b]$.

13 NOTES

(i) The description of I null sets as a set of points of "divergence" of a sequence with summable norms dates back to F. Riesz or earlier.

(ii) The notion of integral used here is due to M. H. Stone, who formulated an abstract version of Daniell's construction of the Lebesgue integral. The role of the truncation axiom needs elucidation. It ensures that the domain of an integral contains "enough characteristic functions" so that the induced measure completely determines the integral (see theorem 5.11).

If the truncation axiom is not assumed, then the methods of this section can be varied to construct an extension of a positive, countably additive linear functional on a vector function lattice which is, modulo the subspace of members of norm zero, a space L with the properties: L is a complete normed space (a **Banach space**), L is a vector lattice such that if $|f| \geqq |g|$ then $\|f\| \geqq \|g\|$, and if $f \geqq 0$ and $g \geqq 0$, then $\|f + g\| = \|f\| + \|g\|$. S. Kakutani calls such spaces ***L* spaces**, and he has established a structure theory for these. See, for example, the appendix of [KN].

(iii) We have not yet defined the "integral I_μ on $L_1(\mu)$ with respect to a measure μ on $\mathscr{A}$". An obvious possibility: since a measure is an exact function that is continuous at $\varnothing$, it induces a pre-integral I^μ on the class of $\mathscr{A}$ simple functions, and one could define the integral I_μ on $L_1(\mu)$ to be the Daniell extension of I^μ. However we shall presently show that there is an integral extension of I^μ that generally has a smaller domain than does the Daniell extension—its null completion is the Daniell extension. The integral I_μ, which we define in chapter 6, turns out to be the minimal integral extension of I^μ.

Chapter 4

INTEGRAL TO MEASURE

A **measure** is a real valued, non-negative, countably additive function on a δ-ring $\mathscr{A}$. A **δ-ring** is a ring $\mathscr{A}$ of sets such that if $\{A_n\}_n$ is a sequence in $\mathscr{A}$. then $\bigcap_n A_n \in \mathscr{A}$; that is, a δ-ring is a ring $\mathscr{A}$ that is closed under countable intersection. The family of all finite subsets of $\mathbb{R}$, the family of all countable subsets of $\mathbb{R}$, and the family of all bounded subsets of $\mathbb{R}$ are examples of δ-rings. We observe that one of these families is closed under countable union but the other two are not.

Here are two examples of a measure. If X is any set, **counting measure** γ for X is defined for all finite subsets A of X by letting $\gamma(A)$ be the number of members of A. If f is a non-negative, real valued function on X, then **discrete measure with weight function** f, v_f, is defined by $v_f(A) = \sum_{x \in A} f(x)$ for all finite subsets A of X.

At this stage we have no assurance that measures exists, except for essentially trivial examples. Generally one must extend a function on some family of sets, such as a length function on the family of intervals, to a measure on some δ-ring containing the family. We will accomplish this extension by showing that each integral induces a measure in a natural way. Then a length function (or pre-measure) induces a pre-integral, which induces an integral, which in turn induces a measure, and this measure is an extension of the length function (pre-measure)!

Suppose J is an integral on M. Let $\mathscr{B} = \{B: \chi_B \in M\}$ and let $\eta(B) = J(\chi_B)$. The family $\mathscr{B}$ is a ring because $\chi_{A \cup B} = \chi_A \vee \chi_B$, $\chi_{A \cap B} = \chi_A \wedge \chi_B$ and $\chi_{A \setminus B} = \chi_A - \chi_{A \cap B}$, and the function η is finitely additive because J is linear. If $\{B_n\}_n$ is an increasing sequence in $\mathscr{B}$, $sup_n\, \eta(B_n) < \infty$ and $B = \bigcup_n B_n$, then $\{\chi_{B_n}\}_n$ is an increasing sequence in M that converges pointwise to χ_B, and $sup_n J(\chi_{B_n}) < \infty$. Consequently, since J is an integral, $\chi_B \in M$ and $lim_n J(\chi_{B_n}) = J(\chi_B)$, so $B \in \mathscr{B}$ and $lim_n \eta(B_n) = \eta(B)$. In particular, η is continuous from below and hence is countably additive. If $\{C_n\}_n$ is a decreasing sequence in $\mathscr{B}$, then $\{C_1 \setminus C_n\}_n$ is increasing, and we infer that $\bigcup_n (C_1 \setminus C_n) = C_1 \setminus \bigcap_n C_n \in \mathscr{B}$, so $\bigcap_n C_n \in \mathscr{B}$ and $\mathscr{B}$ is therefore a δ-ring. Thus η is a measure. It is the **measure induced by the integral** J.

The argument given in the preceding paragraph also shows that the measure η has the special property: if $\{A_n\}_n$ is a disjoint sequence in $\mathscr{B}$ such that $\sum_n \eta(A_n) < \infty$, then $\bigcup_n A_n \in \mathscr{B}$. A measure with this property is called a **standard measure**. Thus the measure induced by an integral is always a standard measure. For the record:

1 Proposition (Integral to Measure) *If J is an integral on M, $\mathscr{B} = \{B: \chi_B \in M\}$ and $\eta(B) = J(\chi_B)$ for B in $\mathscr{B}$, then $\mathscr{B}$ is a δ-ring and η is a standard measure on $\mathscr{B}$.*

We use the preceding proposition to construct an abundance of measures. Each pre-measure μ on $\mathscr{A}$ induces a pre-integral I^μ on $L^{\mathscr{A}}$ according to theorem 2.7, where $L^{\mathscr{A}}$ is the class of $\mathscr{A}$ simple functions and $I^\mu(\chi_A) = \mu(A)$ for A in $\mathscr{A}$. The Daniell extension I^1 on L^1 of the pre-integral I^μ is an integral according to theorem 3.4, and proposition 1 asserts that I^1 induces a measure η on $\mathscr{B} = \{B: \chi_B \in L^1\}$ and $\eta(B) = I^1(\chi_B)$ for $B \in \mathscr{B}$. If $A \in \mathscr{A}$ then $\chi_A \in L^{\mathscr{A}} \subset L^1$ so $A \in \mathscr{B}$, and $\mu(A) = I^\mu(\chi_A) = I^1(\chi_A) = \eta(A)$. Hence η is an extension of μ, and the following theorem is proved.

2 Extension Theorem for Pre-Measures *Each pre-measure μ on $\mathscr{A}$, and in particular each exact function on $\mathscr{A}$ that is continuous at $\varnothing$, can be extended to a measure.*

Explicitly: The measure induced by the Daniell extension of the pre-integral I^μ on $L^{\mathscr{A}}$ is an extension of μ.

There is usually not a unique measure that extends a pre-measure but it turns out that there is a "smallest" such extension. We first observe there is a smallest δ-ring that contains a family $\mathscr{A}$ of sets, because $\mathscr{A}$ is a subfamily of the δ-ring $\{B: B \subset \bigcup_{A \in \mathscr{A}} A\}$, and $\mathscr{D} = \{D$: *D belongs to each δ-ring that contains $\mathscr{A}$*$\}$ is the smallest δ-ring that contains $\mathscr{A}$. The family $\mathscr{D}$ is the **δ-ring generated by** $\mathscr{A}$; it contains $\mathscr{A}$ and is a subfamily of every δ-ring that contains $\mathscr{A}$.

If μ is a pre-measure on $\mathscr{A}$, then the δ-ring $\mathscr{D}$ generated by $\mathscr{A}$ is a subfamily of the domain of every measure η that extends μ, so $\eta | \mathscr{D}$ is an extension of μ. We presently show that two measures that agree on a family that is closed under intersection (as $\mathscr{A}$ is) agree on the generated δ-ring, from which we conclude that there is a *unique* measure ν $(= \eta | \mathscr{D})$

on $\mathscr{D}$ that extends μ. We assume for the moment that this is the case, in order to simplify the statement of the next two results. The unique measure ν on $\mathscr{D}$ that extends μ is the **measure induced by** μ, or the **minimal measure extending** μ; every measure that extends μ also extends ν. In general, ν is not standard.

The δ-ring generated by the family of compact subsets of $\mathbb{R}$ is the **Borel δ-ring $\mathscr{B}^\delta(\mathbb{R})$** for $\mathbb{R}$. The family of bounded subsets is a δ-ring that contains the family of compact sets and hence contains $\mathscr{B}^\delta(\mathbb{R})$. A measure on $\mathscr{B}^\delta(\mathbb{R})$ is a **Borel measure** for $\mathbb{R}$. The family $\mathscr{B}^\delta(\mathbb{R})$ is also generated by the class of closed intervals, and each length function is a pre-measure, so the minimal measure extending a length function is a Borel measure. We show that every Borel measure can be obtained in this way.

3 PROPOSITION *The minimal measure extending a length function is a Borel measure for* $\mathbb{R}$, *and every Borel measure is such an extension.*

PROOF It's only necessary to show that the restriction of a Borel measure μ to the family of closed intervals is a length function. If $a \leqq b \leqq c$ then $\mu[a:b] + \mu[b:c] = \mu[a:b] + \mu(b:c] + \mu\{b\} = \mu[a:c] + \mu\{b\}$, so μ has the required additive property. The boundary inequality, if $a < b$ then $\mu[a:b] \geqq \mu\{a\} + \mu\{b\}$, is equally evident. Finally $lim_n \, \mu[a - n^{-1}: a + n^{-1}] = \mu\{a\}$ because μ is continous from above at $\{a\}$, and it follows that μ satisfies the continuity condition for lengths. ∎

A length function λ on $\mathscr{J}$ is translation invariant iff $\lambda[a:b] = \lambda[a + c:b + c]$ for all a, b and c with $a \leqq b$. According to theorem 1.5, the usual length ℓ, where $\ell[a:b] = b - a$, is the only length function that is translation invariant and assigns the value 1 to $[0:1]$. Consequently, in view of the preceding corollary, there is precisely one Borel measure Λ with the property that $\Lambda([0:1]) = 1$ and $\Lambda([a:b]) = \Lambda([a + c:b + c])$ for all a, b and c with $a \leqq b$. It turns out that this **measure is translation invariant** in the sense that if $c \in \mathbb{R}$, $B \in \mathscr{B}^\delta(\mathbb{R})$ and $c + B = \{c + b: b \in B\}$, then $c + B \in \mathscr{B}^\delta(\mathbb{R})$ and $\Lambda(B) = \Lambda(c + B)$. We leave the proof of this fact to the reader (it follows from the fact that the pre-integral I^λ on $L^{\mathscr{J}}$ is translation invariant in the sense that if $f \in L^{\mathscr{J}}$ and $f_c(x) = f(x - c)$ then $f_c \in L^{\mathscr{J}}$ and $I^\lambda(f) = I^\lambda(f_c)$). Thus:

4 PROPOSITION *There is one and only one translation invariant Borel measure* Λ *such that* $\Lambda([0:1]) = 1$.

The measure Λ is called **Borel–Lebesgue measure for** $\mathbb{R}$. It is not standard since $\mathbb{N} \notin \mathscr{B}^\delta(\mathbb{R})$ although $\{n\} \in \mathscr{B}^\delta(\mathbb{R})$ and $\Lambda(\{n\}) = 0$ for all n in $\mathbb{N}$.

We need some facts about δ-rings before establishing the uniqueness of the minimal measure extending a pre-measure. The union of countably many members of a δ-ring may fail to be a member, but each δ-ring is "locally" closed under countable union, in a sense made precise by the following lemma. We agree that a ring $\mathscr{A}$ is closed under **dominated countable union** iff $\bigcup_n A_n \in \mathscr{A}$ whenever $\{A_n\}_n$ is a sequence in $\mathscr{A}$ and there is a member B of $\mathscr{A}$ such that $B \supset \bigcup_n A_n$.

5 LEMMA *Each δ-ring is closed under dominated countable union, and each ring that is closed under dominated countable disjoint union is a δ-ring.*

PROOF If $\{A_n\}_n$ is a sequence in a δ-ring $\mathscr{A}$ and $\{A_n\}_n$ is dominated by a member B of $\mathscr{A}$, then $B \setminus \bigcup_n A_n = \bigcap_n (B \setminus A_n) \in \mathscr{A}$ and so $\bigcup_n A_n = B \setminus (B \setminus \bigcup_n A_n) \in \mathscr{A}$.

On the other hand, if $\{A_n\}_n$ is a sequence in a ring $\mathscr{A}$ and $B_n = \bigcap_{k \leqq n} A_k$ for each n, then the sequence $\{B_n\}_n$ of partial intersections is decreasing, the difference sequence $\{B_n \setminus B_{n+1}\}_n$ is disjoint and dominated by B_1, and so $\bigcup_n (B_n \setminus B_{n+1}) = B_1 \setminus \bigcap_n B_n = A_1 \setminus \bigcap_n A_n \in \mathscr{A}$. Hence $\bigcap_n A_n \in \mathscr{A}$ provided $\mathscr{A}$ is closed under dominated, countable disjoint union. Consequently, in this case, $\mathscr{A}$ is a δ-ring. ■

6 THEOREM ON GENERATED δ-RINGS *Suppose a family $\mathscr{A}$ is closed under finite intersections. Then the smallest family $\mathscr{C}$ that contains $\mathscr{A}$ and is closed under proper difference, finite disjoint union and dominated countable disjoint union, is the δ-ring generated by $\mathscr{A}$.*

PROOF It is sufficient to show that $\mathscr{C}$ is closed under intersection, for $\mathscr{C}$ is then a ring which, in view of the preceding lemma, is a δ-ring.

For each set B the family $\mathscr{C}_B = \{C\colon C \in \mathscr{C} \text{ and } B \cap C \in \mathscr{C}\}$ is closed under proper difference, disjoint union and dominated countable disjoint union, because $\mathscr{C}$ is closed under these operations and "intersection with B" distributes over each operation (e.g., if $\{D_n\}_n$ is a disjoint sequence in $\mathscr{C}_B$ that is dominated by a member E of $\mathscr{C}_B$, then $\{D_n\}_n$ and $\{B \cap D_n\}_n$ are disjoint sequences in $\mathscr{C}$ and are dominated by $E \in \mathscr{C}$, so $\bigcup_n D_n \in \mathscr{C}$ and $\bigcup_n (B \cap D_n) = B \cap \bigcup_n D_n \in \mathscr{C}$, whence $\bigcup_n D_n \in \mathscr{C}_B$).

If $B \in \mathscr{A}$ then $\mathscr{C}_B \supset \mathscr{A}$ because $\mathscr{C} \supset \mathscr{A}$ and $\mathscr{A}$ is closed under intersection. Thus $B \cap C \in \mathscr{C}$ if $B \in \mathscr{A}$ and $C \in \mathscr{C}$. Hence, if $B \in \mathscr{C}$ then $\mathscr{C}_B \supset \mathscr{A}$, and since $\mathscr{C}_B$ has the three closure properties, $\mathscr{C}_B \supset \mathscr{C}$. Consequently, $B \cap C \in \mathscr{C}$ for all B and C in $\mathscr{C}$. ■

7 COROLLARY *Two measures that agree on a family $\mathscr{A}$ that is closed under intersection also agree on the δ-ring generated by $\mathscr{A}$.*

Consequently each pre-measure on $\mathscr{A}$ extends to a unique measure, the induced measure, on the δ-ring generated by $\mathscr{A}$.

PROOF Suppose μ is a measure on a δ-ring $\mathscr{B}$, ν is a measure on $\mathscr{C}$ and that $\mathscr{E} = \{E: E \in \mathscr{B} \cap \mathscr{C}$ and $\mu(E) = \nu(E)\}$. If D and E belong to $\mathscr{E}$ then $D \setminus E \in \mathscr{B} \cap \mathscr{C}$ because $\mathscr{B} \cap \mathscr{C}$ is a δ-ring, and if $D \supset E$ then $\mu(D \setminus E) = \mu(D) - \mu(E) = \nu(D) - \nu(E) = \nu(D \setminus E)$ so $D \setminus E \in \mathscr{E}$. Similarly, if $D \cap E = \varnothing$ then $D \cup E \in \mathscr{E}$. Finally, if $\{D_n\}_n$ is a disjoint sequence in $\mathscr{E}$, $E \in \mathscr{E}$ and $\bigcup_n D_n \subset E$, then $\bigcup_n D_n \in \mathscr{B} \cap \mathscr{C}$ and $\mu(\bigcup_n D_n) = \sum_n \mu(D_n) = \sum_n \nu(D_n) = \nu(\bigcup_n D_n)$ so $\bigcup_n D_n \in \mathscr{E}$. The preceding theorem then implies that if $\mathscr{E}$ contains a family $\mathscr{A}$ that is closed under intersection then $\mathscr{E}$ contains the δ-ring $\mathscr{A}$ generates. ■

A measure that is obtained from an exact pre-measure has special approximation properties, derived essentially from the inner approximation property of exactness. We deduce this approximation result from an "above and below" approximation for Daniell integrable functions.

We suppose that *I is a pre-integral on L* and that I^1 on L^1 is its Daniell extension.

8 LEMMA (APPROXIMATION FROM ABOVE AND BELOW) *If I is a pre-integral on L, $f \in L^1$ and $e > 0$ then there is a member g of L and a sequence $\{h_n\}_n$ of non-negative members of L such that $\sum_n I(h_n) < e$ and $|f - g|(x) \leqq \sum_n h_n(x)$ for all x.*

Consequently, if $p_n = g - \sum_{i=1}^n h_i$ and $q_n = g + \sum_{i=1}^n h_i$, then $\{p_n\}_n$ is decreasing in L, $\{q_n\}_n$ is increasing in L, $\lim_n p_n \leqq f \leqq \lim_n q_n$, and $0 \leqq \lim_n I(q_n) - \lim_n I(p_n) < e$.

PROOF Since $f \in L^1$, $f = \sum_n f_n$ outside of an I null set E, for some norm summable sequence $\{f_n\}_n$ in L. For $e > 0$ choose N so that $\sum_{n>N} \|f_n\|_1 < e/2$ and choose $\{v_n\}_n$ in L with $\sum_n \|v_n\|_1 < e/2$ and $\sum_n |v_n| = \infty$ on E. Let $g = \sum_{i=1}^N f_i$ and let $h_n = |f_{N+n}| + |v_n|$ for each n. Then g and $\{h_n\}_n$ have the desired properties. ■

Suppose ν is a measure on $\mathscr{C}$. We say that a member C of $\mathscr{C}$ has **inner ν approximations in $\mathscr{E}$** iff $\mathscr{E}$ is a subfamily of $\mathscr{C}$ and $\nu(C) = sup\{\nu(E): E \in \mathscr{E}$ and $E \subset C\}$, and C has **outer approximations in $\mathscr{E}$** iff $\nu(C) = inf\{\nu(E): E \in \mathscr{E}$ and $E \supset C\}$. We agree that for each family $\mathscr{A}$, $\mathscr{A}_\delta$ is the family of intersections of countably many members of $\mathscr{A}$, and $\mathscr{A}_\sigma$ is the family of countable unions of members. Thus $\mathscr{A}_{\delta\sigma}$ consists of countable unions of countable intersections of members of $\mathscr{A}$.

9 PROPOSITION (INNER APPROXIMATION) *If I is the pre-integral induced by an exact pre-measure on $\mathscr{A}$ and ν on $\mathscr{C}$ is the measure induced* *by the Daniell extension I^1 of I, then each member of $\mathscr{C}$ has inner approximations in $\mathscr{A}_\delta$.*

Moreover, for each C in $\mathscr{C}$ there is A in $\mathscr{A}_{\delta\sigma}$ such that $A \subset C$ and $v(C \setminus A) = 0$.

PROOF We first show that if $\{C_n\}_n$ is decreasing sequence in $\mathscr{C}$ and each C_n has inner approximations in $\mathscr{A}$ then $C = \bigcap_n C_n$ has such approximations in $\mathscr{A}_\delta$. Choose $A_n \in \mathscr{A}$ so that $A_n \subset C_n$ and $\sum_n v(C_n \setminus A_n)$ is small. Then $A = \bigcap_n A_n$ is an inner approximation for C because $C \setminus A \subset \bigcup_n (C_n \setminus A_n)$, whence, by countable subadditivity of v, $v(C \setminus A) \leqq \sum_n v(C_n \setminus A_n)$.

Next, for each C in $\mathscr{C}$ and each $e > 0$ there is, by the approximation lemma, a decreasing sequence $\{g_n\}_n$ of $\mathscr{A}$ simple functions with $lim_n g_n \leqq \chi_C$ and $lim_n I(g_n) \geqq v(C) - e$. We may suppose that $g_n \geqq 0$ for each n, so $0 \leqq lim_n g_n \leqq \chi_C$. If $0 < a < 1$ and $E = \{x\colon lim_n g_n(x) \geqq a\}$, then $E \subset C$ and $lim_n g_n \leqq \chi_E + a\chi_{C \setminus E}$, so $I^1(lim_n g_n) \leqq v(E) + av(C \setminus E) = (1 - a)v(E) + av(C)$ provided $E \in \mathscr{C}$. On the other hand $I^1(lim_n g_n) = lim_n I(g_n) \geqq v(C) - e$, so $v(C) - e \leqq (1 - a)v(E) + av(C)$, whence $v(C) \leqq v(E) + (e/(1 - a))$, so E is an inner approximation for C, provided $E \in \mathscr{C}$.

Finally, $E = \bigcap_n E_n$ where $E_n = \{x\colon g_n(x) \geqq a\}$, so $E \in \mathscr{C}$, and in view of the result established in the first paragraph, the proof reduces to showing that there is an inner v approximation for E_n in $\mathscr{A}$ for each n. But g_n is $\mathscr{A}$-simple so E_n belongs to the ring generated by the lattice $\mathscr{A}$, which consists of unions of disjoint differences of members of $\mathscr{A}$ (see theorem 2.2). The definition of exactness then shows that there is an inner approximation for E_n in $\mathscr{A}$, and the first half of proposition follows.

If $C \in \mathscr{C}$ and $A_n \in \mathscr{A}_\delta$ with $v(C \setminus A_n) < 1/n$ for each n, then $\bigcup_n A_n \in \mathscr{A}_{\delta\sigma}$ and $v(C \setminus \bigcup_n A_n) = 0$. The proposition is thus established. ∎

The preceding proposition implies results on outer approximation, based on the observation that if $C \subset D$ and E is a "good" inner approximation for $D \setminus C$, then $D \setminus E$ is a "good" outer approximation for C. The proof of corollary 10 will furnish an example.

A measure μ is a **Lebesgue–Stieltjes measure for** $\mathbb{R}$ iff there is a length function λ such that μ is induced by the Daniell extension of the pre-integral I^λ. We say that μ is the **Lebesgue–Stieltjes measure induced by** λ. We notice that μ determines λ—in fact, $\lambda = \mu | \mathscr{J}$, where $\mathscr{J}$ is the class of closed intervals—so length functions and Lebesgue–Stieltjes measures are paired off. It is worth noticing that the domain of a Lebesgue–Stieltjes measure μ depends on the length function $\mu | \mathscr{J}$, whereas all minimal measures induced by length functions have the same domain, the Borel δ-ring $\mathscr{B}^\delta(\mathbb{R})$.

Lebesgue measure Λ^1 for $\mathbb{R}$ is the Lebesgue–Stieltjes measure induced by the usual length ℓ. It is straightforward to verify that Λ^1 is translation invariant and that $\Lambda^1([0:1]) = 1$. If μ is a translation invariant Lebesgue–Stieltjes measure and $\mu([0:1]) = 1$, then the length $\lambda = \mu|\mathscr{J}$ is translation invariant and $\lambda([0:1]) = 1$, so $\lambda = \ell$ by theorem 1.5. Consequently each of μ and Λ^1 is induced by the usual length ℓ, and it follows that Λ^1 is the unique Lebesgue–Stieltjes measure that is translation invariant and assigns the value 1 to $[0:1]$.

Proposition 9 has the following corollary about Lebesgue–Stieltjes measures.

10 COROLLARY *If B belongs to the domain $\mathscr{S}$ of a Lebesgue Stieltjes measure η, and in particular if $B \in \mathscr{B}^\delta(\mathbb{R})$, then for $e > 0$ there is a compact set K and an open set U such that $K \subset B \subset U$ and $\eta(K) + e \geqq \eta(B) \geqq \eta(U) - e$.*

PROOF The measure η is induced by the Daniell extension of I^λ where $\lambda = \eta|\mathscr{J}$, so by proposition 9, B has inner approximations in $\mathscr{A}_\delta$, where $\mathscr{A}$ is the lattice generated by the family of closed intervals. But $\mathscr{A}_\delta$ is then just the family of compacta, so there is a compact set K with $K \subset B$ and $\eta(K) + e \geqq \eta(B)$.

If B is a bounded set, then $B \subset (a:b)$ for some a and b. There is then a compact subset D of $(a:b)\setminus B$ such that $\eta(D) + e \geqq \eta((a:b)\setminus B) = \eta((a:b)) - \eta(B)$. Then $\eta(B) + e \geqq \eta((a:b)) - \eta(D)$, and if $U = (a:b)\setminus D$, then $B \subset U$ and U is an open set as required.

If B is not bounded it is the union of the bounded members $B_n = B \cap [-n:n]$ of $\mathscr{S}$, there is an open set $U_n \supset B_n$ so that $\eta(U_n) < \eta(B) + 2^{-n}$, and $U = \bigcup_n U_n$ is an open set as required. ■

The preceding corollary is often phrased: each Lebesgue–Stieltjes measure for $\mathbb{R}$, and each Borel measure for $\mathbb{R}$, is **inner regular** (inner approximation by compacta) and **outer regular** (outer approximation by open sets); in other words each such measure is **regular**.

The minimal measure induced by a pre-measure λ on $\mathscr{A}$ is generally *not* the measure induced by the Daniell extension of the pre-integral I^λ on $L^{\mathscr{A}}$. Each measure induced by an integral is standard, and the minimal measure may fail to be standard (e.g., Borel–Lebesgue measure Λ). Moreover, each Daniell integral is null complete (each function with null support is integrable), and the measure ν on $\mathscr{C}$ induced by a null complete integral has the special property: if $E \subset D \in \mathscr{C}$ and $\nu(D) = 0$, then $E \in \mathscr{C}$ and $\nu(E) = 0$. Such measures are called **complete**. Borel–Lebesgue measure is *not* complete. However, it turns out that the minimal measure induced by a pre-measure λ completely determines the measure induced by the Daniell extension of the pre-integral I^λ.

The **standardization** of a measure μ on $\mathscr{A}$ is the function ν on $\mathscr{B}$ given by: $B \in \mathscr{B}$ iff $B = \bigcup_n A_n$ for some disjoint sequence $\{A_n\}_n$ in $\mathscr{A}$ with $\sum_n \mu(A_n) < \infty$, and in this case $\nu(B) = \sum_n \mu(A_n)$. We leave to the reader the proof that this definition is not ambiguous, that ν is a standard measure; that μ is standard iff it is identical with its standardization, and that μ is complete iff its standardization ν is complete.

The **usual completion** or just the **completion** of a measure μ on $\mathscr{A}$ is the function $\mu^\vee$ on $\mathscr{A}^\vee$ defined by: $A^\vee \in \mathscr{A}^\vee$ iff $A^\vee$ is the symmetric difference $A \triangle E$ for some member A of $\mathscr{A}$ and some subset E of a member of $\mathscr{A}$ of μ measure zero, and in this case $\mu^\vee(A^\vee) = \mu(A)$. (Roughly speaking, μ is defined on the sets which are, give and take a subset of a set of measure zero, members of $\mathscr{A}$.) It is not hard to verify that $\mu^\vee$ is well defined, it is a complete measure extending μ, and it is the smallest such extension.

The standardization of the completion of a measure μ is its **standardized completion**. It is an extension of μ that is both standard and complete, and it is the smallest such extension.

11 PROPOSITION *Let λ be a pre-measure on $\mathscr{A}$, I^λ on $L^{\mathscr{A}}$ the induced pre-integral and let ν on $\mathscr{C}$ be the measure induced by the Daniell extension of I^λ. Then every complete standard measure that extends λ is an extension of ν.*

Consequently ν is the standardized completion of the minimal measure extending λ.

PROOF Let $L^1 = (L^{\mathscr{A}})^1$ be the domain of the Daniell extension of I^λ, and let σ on $\mathscr{D}$ be the standardized completion of the minimal measure extending λ. Since ν is standard and complete, ν is an extension of σ. Thus $\mathscr{A} \subset \mathscr{D} \subset \mathscr{C}$, $L^{\mathscr{A}} \subset L^{\mathscr{D}} \subset L^{\mathscr{C}} \subset L^1$, and consequently the Daniell extensions, $(I^\lambda)^1$ and $(I^\sigma)^1$, of the pre-integrals are ordered similarly. Thus $(L^{\mathscr{A}})^1 \subset (L^{\mathscr{D}})^1 \subset L^1 = (L^{\mathscr{A}})^1$ whence I^λ and I^σ have the same Daniell extensions. In particular, ν is the measure induced by the Daniell extension of I^σ, so proposition 9 applies to the pre-integral I^σ and the measure ν. An application of the same yields that σ is an extension of ν for the following reason: Suppose $C \in \mathscr{C}$. Since $\chi_C \in L^1 = (L^{\mathscr{D}})^1$ and σ on $\mathscr{D}$ is standard, it is clear that C is a subset of a member D of $\mathscr{D}$. Applying proposition 9 to the members C and $D \setminus C$ of $\mathscr{C}$ we find sets $F \subset C$ and $G \subset D \setminus C$ in $\mathscr{D}$ so that $\nu(C \setminus F) = 0$ and $\nu((D \setminus C) \setminus G) = 0$. Thus C is sandwiched between the members F and $D \setminus G$ of $\mathscr{D}$, $\sigma(F) = \nu(C) = \sigma(D \setminus G)$, and σ is a complete measure. Consequently $C \in \mathscr{D}$. ■

12 COROLLARY *Each Lebesgue–Stieltjes measure ν for $\mathbb{R}$ is the standardized completion of the Borel measure $\nu|\mathscr{B}^\delta(\mathbb{R})$.*

In particular, Lebesgue measure for $\mathbb{R}$ is the standardized completion of Borel–Lebesgue measure.

13 NOTES (i) Lebesgue measure Λ^1 on $\mathscr{L}$ is the measure induced by the Daniell extension of I^ℓ, where ℓ is the length, so Λ^1 is the minimal complete standardized measure that extends ℓ. The family $\mathscr{L}_\sigma$, consisting of all countable unions of members of $\mathscr{L}$, is the smallest σ-ring (a ring closed under countable unions) containing $\mathscr{L}$, and **Classical Lebesgue measure** is the extension of Λ^1 to $\mathscr{L}_\sigma$ given by assigning infinite measure to members of $\mathscr{L}_\sigma \setminus \mathscr{L}$. The σ-ring $\mathscr{L}_\sigma$ is in fact a σ-field (closed under complements and countable unions), and its members are called **Lebesgue measurable sets**. Similarly, the members of the smallest σ-field $(\mathscr{B}^\delta(\mathbb{R}))_\sigma$ containing $\mathscr{B}^\delta(\mathbb{R})$ are called **Borel measurable**.

(ii) There are translation invariant measures on σ-fields for $\mathbb{R}$ which are extensions of classical Lebesgue measure (see S. Kakutani and J. C. Oxtoby, *Ann. of Math.* **52** (1950) 580–590). But there is no measure extending Lebesgue measure which is defined for *all* subsets of $\mathbb{R}$, even without the requirement of translation invariance. However Banach showed that there are non-negative *finitely additive* extensions of Lebesgue measure which are defined for all subsets of $\mathbb{R}$, which are invariant under rigid motions (*Fund. Math.* **4** (1923), 7–33).

(iii) Borel–Lebesgue measure Λ_n for $\mathbb{R}^n$ can be defined by imitating the construction in $\mathbb{R}$ (see Supplement) or by recursively setting Λ_{n+1} equal to the product measure $\Lambda_n \otimes \Lambda$ as soon as we have defined product measure (chapter 7). Then **Lebesgue measure Λ^n** is the standardized completion of Λ_n. It is translation invariant (and invariant under all rigid motions) for each n. There are *finitely additive* extensions, invariant under rigid motions, of Λ^2 to the family of all subsets of $\mathbb{R}^2$ (see Banach, *loc. cit.*). But for $n \geqq 3$, such extensions of Λ^n do not exist (see Banach and Tarski, *Fund. Math.* **6** (1924), 244–277; see also J. Rosenblatt, *Trans. Amer. Math. Soc.* **265** (1981), 623–636).

(iv) The measure Λ^n is the unique measure on its domain, up to constant multiple, that is invariant under rigid motions. In fact (G. A. Margulis, *Monatsh. Math.* **90** (1980), 233–235 and preprint 1981; D. Sullivan, *Bull. Amer. Math. Soc.* (*N.S.*) **4** (1981), 121–123) for $n \geqq 3$, there is *no finitely additive* non-negative function on the family of Lebesgue measurable sets in $\mathbb{R}^n$, other than a multiple of Lebesgue measure, which is invariant under rigid motions. The conclusion extends to $\mathbb{R}^2$ provided we require invariance under the shear transformation: $(x_1, x_2) \mapsto (x_1 + x_2, x_2)$, as well as invariance under rigid motions (S. Wagon, *Proc. Amer. Math. Soc.* **85** (1982), 353–359).

SUPPLEMENT: LEBESGUE MEASURE Λ^n FOR $\mathbb{R}^n$

Lebesgue measure for $\mathbb{R}^n$, $n > 1$, is the natural generalization of Lebesgue measure for $\mathbb{R}$. Recall that a closed interval in $\mathbb{R}^n$ is the Cartesian product $\times_{i=1}^n A_i$ of closed intervals $\{A_i\}_{i=1}^n$ in $\mathbb{R}$ and its volume $\lambda_n(\times_{i=1}^n A_i)$ is $\prod_{i=1}^n \ell(A_i)$, the product of the lengths of its sides. The volume function λ_n on the class $\mathscr{I}_n$ of closed intervals induces, according to proposition 2.10, a pre-integral I^n on the class of $\mathscr{I}_n$ simple functions by the prescription: $I^n(\sum_{i=1}^k a_i \chi_{A_i}) = \sum_{i=1}^k a_i \lambda_n(A_i)$ for a_i in $\mathbb{R}$ and A_i in $\mathscr{I}_n$, $i = 1, 2, \ldots k$. The **Lebesgue integral** for $\mathbb{R}^n$ is the Daniell extension of I^n and Lebesgue measure for $\mathbb{R}^n$, Λ^n on $\mathscr{L}^n$, is the measure induced by the Lebesgue integral.

The characteristic function of each closed interval in $\mathbb{R}^n$ belongs to the domain of the Lebesgue integral, and consequently the δ-ring $\mathscr{B}^\delta(\mathbb{R}^n)$ generated by the family of closed intervals is a subfamily of $\mathscr{L}^n$. **Borel–Lebesgue measure Λ_n** for $\mathbb{R}^n$ is $\Lambda^n|\mathscr{B}^\delta(\mathbb{R}^n)$. We list, omitting the proofs, some straightforward generalizations of the facts about Lebesgue measure Λ^1.

Recall that the iterated Riemann integral on $C_c(\mathbb{R}^n)$ is a pre-integral according to proposition 2.11.

14 PROPOSITION *Borel–Lebesgue measure Λ_n is the minimal measure that is an extension of the volume function, Λ^n is the standardized completion of Λ_n, and both Λ_n and Λ^n are translation invariant.*

The Daniell extension of the iterated Riemann integral on $C_c(\mathbb{R}^n)$ is identical with the Lebesgue integral.

Each member of $\mathscr{L}^n$ has compact inner and open outer approximations, so both Λ^n and Λ_n are regular measures.

Classical Lebesgue measure for $\mathbb{R}^n$ is the extension of Λ^n to the σ-ring $\mathscr{L}_\sigma^n$ generated by $\mathscr{L}^n$, given by assigning infinite measure to members of $\mathscr{L}_\sigma^n \setminus \mathscr{L}^n$, and the members of $\mathscr{L}_\sigma^n$ are called **Lebesgue measurable sets**.

SUPPLEMENT: MEASURES ON $\mathscr{B}^\delta(X)$

Let us suppose that **X is a locally compact Hausdorff space** and agree that **$\mathscr{B}^\delta(X)$** is the δ-ring generated by the family $\mathscr{C}$ of compact subsets of X. We recall that a content λ for X is a non-negative, real valued, monotonic, additive and subadditive function on $\mathscr{C}$, and that a content λ is regular iff $\lambda(C) = inf\{\lambda(D): D$ a compact neighborhood of $C\}$.

Each regular content λ is exact (proposition 1.9), hence it is an exact pre-measure, and corollary 7 then asserts that there is a unique **Borel measure μ** (that is, a measure on $\mathscr{B}^\delta(X)$) that extends λ. The measure μ is minimal (i.e., each measure that extends λ also extends μ).

15 THEOREM *Each regular content λ for X can be extended to a unique Borel measure μ, and every measure that extends λ also extends μ.*

Moreover, the measure μ is regular, and if $B \in \mathscr{B}^\delta(X)$ there is D in $\mathscr{C}_\sigma$ such that $D \subset B$ and $\mu(B \setminus D) = 0$.

The last assertion of the theorem's statement is a direct consequence of proposition 9. We use the term **regular** for a measure μ on a family of subsets of X to mean both **inner regular** (inner μ approximations by compacta) and **outer regular** (outer μ approximations by open sets). An inner regular measure μ on $\mathscr{B}^\delta(X)$ is automatically outer regular because: if $B \in \mathscr{B}^\delta(X)$, V is an open member of $\mathscr{B}^\delta(X)$ that contains B and K is a compact set that is a good inner μ approximation to $V \setminus B$, then $V \setminus K$ is an open set that is a good outer μ approximation for B.

There are Borel measures that are *not* regular—an example is not hard to describe. Let X be the set of all ordinal numbers less than or equal to the first uncountable ordinal Ω, and let X have the order topology (the family of sets of the form $\{x: a < x < b\}$ for a and b in X, is a base). Then X is a compact Hausdorff space. The space $X \setminus \{\Omega\}$ has the curious property: any two closed uncountable subsets of $X \setminus \{\Omega\}$ intersect and the intersection is a closed uncountable set (see the interlacing lemma [K] p. 131). Moreover, the intersection of a decreasing sequence $\{F_n\}_n$ of closed uncountable subsets of $X \setminus \{\Omega\}$ is a set of the same kind because: for each $\beta < \Omega$ there is α_n in F_n with $\alpha_n > \beta$ so $\{\alpha_n\}$ has a cluster point $\gamma \geqq \beta$ and $\gamma \in \bigcap_n F_n$. For B in $\mathscr{B}^\delta(X)$, define $\mu(B) = 1$ iff B contains a closed uncountable subset of $X \setminus \{\Omega\}$, and let $\mu(B) = 0$ otherwise. Since there are not two disjoint members of $\mathscr{B}^\delta(X)$ at which μ is non-zero, μ is additive. If $\{A_n\}_n$ is decreasing and $\mu(A_n\} = 1$ for each n, then $\mu(\bigcap_n A_n) = 1$, and it follows that μ is a measure. It is not regular since $\mu(X \setminus \{\Omega\}) = 1$ but μ vanishes at each compact subset of the open set $X \setminus \{\Omega\}$. For each a in X the sets $X_a = \{x: a \leqq x\}$ are compact and $\bigcap_a X_a = \{\Omega\}$. Evidently $\mu(X_a) = 1$ for $a < \Omega$ and $\mu\{\Omega\} = 0$. Consequently the set $\{\Omega\}$ has no outer approximation by open sets, μ is not hypercontinuous on the family $\mathscr{C}$ of compact sets and $\mu|\mathscr{C}$ is a content which is not regular.

SUPPLEMENT: G INVARIANT MEASURES

Let us suppose that G is a group, that a locally compact Hausdorff space X is a left G space, and that for each a in G the map $x \mapsto ax$ is a homeomorphism of X onto itself. The action of G is **transitive** iff for x and y in X there is a in G so $ax = y$, and the action is **semi-rigid** iff for disjoint compact subsets A and B of X and for x in X, there is a neighborhood V of x so no set of the form aV intersects both A and B.

Theorem 1.12 asserts that if the action of G is transitive and semi-rigid then there is a G invariant regular content λ for X that is not identically zero. We leave to the reader (consider theorem 6) the verification of the fact that the unique Borel measure μ that extends λ is also G invariant (explicitly, if $B \in \mathscr{B}^\delta(X)$ and $a \in G$ then $aB \in \mathscr{B}^\delta(X)$ and $\mu(B) = \mu(aB)$). Thus:

16 THEOREM *If a group G acts on a locally compact Hausdorff space X by homeomorphisms, and if the action is transitive and semi-rigid, then there is a G invariant Borel measure for X that is not identically zero.*

According to proposition 1.13, the action by left translation of a locally compact topological Hausdorff group upon itself is semi-rigid. Consequently there is a regular Borel measure η, not identically zero, that is **left invariant** in the sense that $\eta(xB) = \eta(B)$ for all B in $\mathscr{B}^\delta(X)$ and x in X. Thus:

17 COROLLARY (EXISTENCE OF HAAR MEASURE) *There is a regular left invariant Borel measure, not identically zero, for each locally compact Hausdorff topological group.*

Such a left invariant measure η is called a **left Haar** measure. A **right invariant Haar measure** is a regular Borel measure ρ not identically zero, such that $\rho(Bx) = \rho(B)$ for x in X and B in $\mathscr{B}^\delta(X)$. If η is a left Haar measure and we set $\rho(E) = \eta(E^{-1})$ then ρ is a right Haar measure, the **right Haar measure corresponding to η**.

Chapter 5

MEASURABILITY AND σ-SIMPLICITY

We need further information on the structure of integrable functions if our theory of integration is to be conveniently usable. For example, if J on M is the Daniell extension of the pre-integral induced by a length function, must every continuous function with compact support belong to M? The answer is not self-evident, although it had certainly better be "yes"! We shall presently find criteria for integrability involving a set theoretic (measurability) requirement, and a magnitude requirement.

Measurable functions play the role in the theory of measure and integration that is played by continuous functions in general topology. We try to emphasize the similarities. After a couple of preliminary results we define a borel space—the analogue of a topological space—and establish a few general propositions about measurable maps. These are then applied to the class of measurable real valued functions.

A **σ-ring** is a ring $\mathscr{A}$ of sets that is closed under countable union; i.e., if $\{A_n\}_n$ is a sequence in $\mathscr{A}$ then $\bigcup_n A_n \in \mathscr{A}$. Each σ-ring $\mathscr{A}$ is a δ-ring, in view of the identity: if $X = \bigcup_n A_n$, then $\bigcap_n A_n = X \setminus (X \setminus \bigcap_n A_n) = X \setminus \bigcup_n (X \setminus A_n)$. On the other hand, the family of finite subsets of any infinite set and the family of bounded subsets of $\mathbb{R}$ are δ-rings that are not σ-rings.

Recall that a field of sets for X is a ring $\mathscr{A}$ such that $X = \bigcup_{A \in \mathscr{A}} A$ and $X \in \mathscr{A}$. A **σ-field**, or **σ-algebra**, or **borel field**, is a field that is closed under countable union or, equivalently, closed under countable intersection. The family of all subsets A of $\mathbb{R}$ such that A or $\mathbb{R} \setminus A$ is countable is a σ-field for $\mathbb{R}$.

The **σ-ring generated** by a family $\mathscr{A}$ of sets is the smallest σ-ring $\mathscr{S}$ containing $\mathscr{A}$. Explicitly, $\mathscr{S} = \{S : S$ *belongs to each σ-ring $\mathscr{B}$ that con-*

tains $\mathscr{A}\}$. The **σ-field generated by $\mathscr{A}$** is, similarly, the smallest σ-field containing $\mathscr{A}$.

Here are some useful elementary facts about generated families. We agree that for each family $\mathscr{A}$, the family $\mathscr{A}_\sigma = \{E : E = \bigcup_n A_n$ *for some* $\{A_n\}_n$ *in* $\mathscr{A}\}$.

1 PROPOSITION *Suppose $\mathscr{A}$ is a non-empty family of subsets of X and that $\mathscr{D}$, $\mathscr{S}$ and $\mathscr{F}$ are respectively the δ-ring, the σ-ring and the σ-field for X generated by $\mathscr{A}$. Then*

(i) *each member of $\mathscr{D}$ (of $\mathscr{S}$) can be covered by finitely (countably, respectively) many members of $\mathscr{A}$*

(ii) *each member of $\mathscr{D}$ (of $\mathscr{S}$) belongs to the δ-ring (σ-ring, respectively) generated by a countable subfamily of $\mathscr{A}$, and*

(iii) *$\mathscr{S} = \mathscr{D}_\sigma$, $\mathscr{D}_\sigma$ is identical with the family of all unions of countable disjoint subfamilies of $\mathscr{D}$, and $F \in \mathscr{F}$ iff either F or $X \setminus F$ belongs to $\mathscr{S}$.*

PROOF We prove only three of the assertions, leaving other similar proofs to the reader.

Let $\mathscr{B} = \{B$: *there is a sequence* $\{A_n\}_n$ *in* $\mathscr{A}$ *with* $B \subset \bigcup_n A_n\}$. One verifies without difficulty that $\mathscr{B}$ is a σ-ring and evidently $\mathscr{A} \subset \mathscr{B}$. Hence $\mathscr{B}$ contains the smallest σ-ring $\mathscr{S}$ that contains $\mathscr{A}$, so if $S \in \mathscr{S}$ then $S \in \mathscr{B}$. Thus each member of $\mathscr{S}$ can be covered by countably many members of $\mathscr{A}$.

Let $\mathscr{E} = \{E$: *there is a countable subfamily* $\mathscr{A}_E$ *of* $\mathscr{A}$ *such that* E *belongs to the δ-ring generated by* $\mathscr{A}_E\}$. If $\{E_n\}_n$ is a sequence in $\mathscr{E}$, then $\bigcup_n \mathscr{A}_{E_n}$ is a countable subfamily of $\mathscr{A}$, the δ-ring $\mathscr{G}$ generated by $\bigcup_n \mathscr{A}_{E_n}$ contains E_n for each n and hence $\bigcap_n E_n \in \mathscr{G} \subset \mathscr{E}$. It follows that $\mathscr{D} \subset \mathscr{E}$, so each member of $\mathscr{D}$ belongs to the δ-ring generated by a countable subfamily of $\mathscr{A}$.

Since $\mathscr{S}$ is a δ-ring, $\mathscr{D} \subset \mathscr{S}$, and so the union of countably many members of $\mathscr{D}$ belongs to $\mathscr{S}$. Let $\mathscr{U} = \{\bigcup_n D_n : \{D_n\}_n$ *is a disjoint sequence in* $\mathscr{D}\}$. Then $\mathscr{D} \subset \mathscr{U} \subset \mathscr{S}$, and we need only show that $\mathscr{U}$ is a σ-ring to conclude that $\mathscr{U} = \mathscr{S}$. The union $\bigcup_n E_n$ of the members of a sequence $\{E_n\}_n$ is also the union of the members of the disjoint sequence $\{D_n\}_n$ where $D_n = E_n \setminus \bigcup_{k<n} E_k$. Consequently $\mathscr{U} = \mathscr{D}_\sigma$ and is therefore closed under countable union. It remains to show that $\mathscr{U}$ is closed under difference. If $\{A_m\}_m$ and $\{B_n\}_n$ are in $\mathscr{D}$ then $\bigcup_m A_m \setminus \bigcup_n B_n = \bigcup_m (A_m \setminus \bigcup_n B_n) = \bigcup_m \bigcap_n (A_m \setminus B_n)$ and $\bigcap_n (A_m \setminus B_n) \in \mathscr{D}$ for each m, and it follows that $\mathscr{U}$ is closed under difference. ■

One of the devices used in the preceding proof is worth noticing. If $\{A_n\}_n$ is a sequence of sets, the **disjointing process** yields a disjoint sequence $\{B_n\}_n$, where $B_n = A_n \setminus \bigcup_{k<n} A_k$, such that $\bigcup_n A_n = \bigcup_n B_n$ and $\bigcup_{k=1}^m A_k = \bigcup_{k=1}^m B_k$ for each m.

The most important σ-field is the **Borel field of** $\mathbb{R}$, denoted $\mathscr{B}(\mathbb{R})$. It is defined to be the σ-field for $\mathbb{R}$ that is generated by the family of all open subsets of $\mathbb{R}$. It is also the σ-field generated by the closed sets, or the compact sets, or the closed intervals $[a:b]$, or the open intervals or by $\mathscr{B}^\delta(\mathbb{R})$. The Borel field of $\mathbb{R}$ is also the σ-ring generated by any one of these families of sets—this σ-ring is, in fact, a σ-field.

The **Borel field** $\mathscr{B}(\mathbb{R}^*)$ **of** $\mathbb{R}^*$, the extended set of real numbers, is the σ-field generated by the family of all open subsets of $\mathbb{R}^*$. Of course there are other descriptions of $\mathscr{B}(\mathbb{R}^*)$ and $\mathscr{B}(\mathbb{R})$. For example, $A \in \mathscr{B}(\mathbb{R}^*)$ iff $A \cap \mathbb{R} \in \mathscr{B}(\mathbb{R})$, and $\mathscr{B}(\mathbb{R}) = (\mathscr{B}^\delta(\mathbb{R}))_\sigma$, the family of countable unions of members of the Borel δ-ring (the δ-ring generated by the compact sets).

Here is a description of $\mathscr{B}(\mathbb{R})$ and of $\mathscr{B}(\mathbb{R}^*)$ that we will need presently.

2 LEMMA *The Borel field $\mathscr{B}(\mathbb{R})$ is the σ-field generated by sets of the form $(-\infty:r)$ with r rational (or alternatively, of the form $(-\infty:r]$, or $(r:\infty)$, or $[r:\infty)$).*

The Borel field $\mathscr{B}(\mathbb{R}^*)$ is similarly generated by sets $[-\infty:r)$ (or $[-\infty:r]$, or $(r:\infty]$, or $[r:\infty]$).

PROOF We prove only that the σ-ring $\mathscr{A}$ generated by $\{(-\infty:r):r$ *rational*$\}$ generates the Borel field $\mathscr{B}(\mathbb{R})$ of $\mathbb{R}$, leaving the remaining arguments to the reader. First, $\mathbb{R} = \bigcup\{(-\infty:r):r \text{ rational}\}$ so $\mathbb{R} \in \mathscr{A}$ and hence $\mathbb{R}\backslash(-\infty:r) = [r:\infty) \in \mathscr{A}$ for each rational r. Then $(s:\infty) = \bigcup\{[r:\infty):r \text{ rational and } r > s\} \in \mathscr{A}$ for every s. Hence each open interval $(s:r)$, with r and s rational, belongs to $\mathscr{A}$, and since each open set is the union of countably many such intervals, each open set belongs to $\mathscr{A}$. Consequently $\mathscr{A}$ contains the Borel field $\mathscr{B}(\mathbb{R})$, and since $\mathscr{A}$ is generated by a subfamily of $\mathscr{B}(\mathbb{R})$, $\mathscr{A} \subset \mathscr{B}(\mathbb{R})$ so $\mathscr{A} = \mathscr{B}(\mathbb{R})$. ■

A **borel space** is an ordered pair $(X, \mathscr{A})$ such that $\mathscr{A}$ is a σ-field of sets for X (in particular, $X = \bigcup_{A \in \mathscr{A}} A$). The members of $\mathscr{A}$ are $\mathscr{A}$ **measurable**, or $\mathscr{A}$ **borel measurable**, or if confusion is unlikely, just **measurable** or **borel**.

A function f on X to Y is $\mathscr{A}-\mathscr{B}$ **measurable** or $\mathscr{A}-\mathscr{B}$ **borel measurable** iff $\mathscr{A}$ is a σ-field for X, $\mathscr{B}$ is a σ-field for Y, and $f^{-1}[B] \in \mathscr{A}$ for each member B of $\mathscr{B}$.

The elementary properties of measurability all follow from three simple remarks, each a precise analogue in statement and proof of a result about continuous functions.

3 PROPOSITION *Suppose $(X, \mathscr{A})$, $(Y, \mathscr{B})$ and $(Z, \mathscr{C})$ are borel spaces, $f: X \to Y$ and $g: Y \to Z$.*

(i) *The family $\mathscr{Q} = \{E: f^{-1}[E] \in \mathscr{A}\}$ is a σ-field and f is $\mathscr{A}-\mathscr{B}$ measurable iff $\mathscr{B} \subset \mathscr{Q}$.*

(ii) *The family* $\mathscr{T} = \{f^{-1}[B]: B \in \mathscr{B}\}$ *is a* σ*-field and* f *is* $\mathscr{A} - \mathscr{B}$ *measurable iff* $\mathscr{A} \supset \mathscr{T}$.

(iii) *If* f *is* $\mathscr{A} - \mathscr{B}$ *measurable and* g *is* $\mathscr{B} - \mathscr{C}$ *measurable, then* $g \circ f$ *is* $\mathscr{A} - \mathscr{C}$ *measurable.*

Thus $\mathscr{Q}$ is the largest σ-field for Y that makes f $\mathscr{A} - \mathscr{Q}$ measurable, and $\mathscr{T}$ is the smallest σ-field for X that makes f $\mathscr{T} - \mathscr{B}$ measurable.

Here is a particular case of part (i) of the preceding proposition. Suppose $f^{-1}[G] \in \mathscr{A}$ for all members G of a family $\mathscr{G}$ of sets that generates $\mathscr{B}$. Then $\mathscr{G}$, and hence $\mathscr{B}$, are subfamilies of $\mathscr{Q}$, and so f is $\mathscr{A} - \mathscr{B}$ measurable. For example, if f is a continuous real valued function on $\mathbb{R}$ then $f^{-1}[V]$ is open for each set V, and such open sets V generate the Borel σ-field $\mathscr{B}(\mathbb{R})$, so f is $\mathscr{B}(\mathbb{R}) - \mathscr{B}(\mathbb{R})$ measurable. If $f: \mathbb{R}^m \to \mathbb{R}^n$, and the **Borel σ-field $\mathscr{B}(\mathbb{R}^n)$** is defined to be the σ-field that is generated by the family of open subsets of $\mathbb{R}^n$, then the same argument shows that f is $\mathscr{B}(\mathbb{R}^m) - \mathscr{B}(\mathbb{R}^n)$ measurable provided f is continuous.

4 COROLLARY *If* $f^{-1}[G] \in \mathscr{A}$ *for all members* G *of a family* $\mathscr{G}$ *that generates* $\mathscr{B}$, *then* f *is* $\mathscr{A} - \mathscr{B}$ *measurable.*

In particular, a continuous function $f: \mathbb{R}^m \to \mathbb{R}^n$ *is* $\mathscr{B}(\mathbb{R}^m) - \mathscr{B}(\mathbb{R}^n)$ *measurable, and* $f: X \to \mathbb{R}^*$ *is* $\mathscr{A} - \mathscr{B}(\mathbb{R}^*)$ *measurable iff* $\{x: f(x) > r\} \in \mathscr{A}$ *for each rational number* r *(alternatively,* $f(x) \geqq r$, $f(x) < r$, *or* $f(x) \leqq r$*).*

PROOF The last statement follows from the fact (lemma 2) that $\mathscr{B}(\mathbb{R}^*)$ is generated by each of the families of sets $\{x: x > r\}$, $\{x: x \geqq r\}$, $\{x: x < r\}$ and $\{x: x \leqq r\}$, for r rational. ■

Suppose that $(X, \mathscr{A})$ is a borel space and that W is a subset of X. The **relativization of $\mathscr{A}$ to W, $\mathscr{A} \| W$,** is $\{A \cap W : A \in \mathscr{A}\}$. The family $\mathscr{A} \| W$ is a σ-field for W, and in fact, $\mathscr{A} \| W$ is the smallest σ-field that makes the identity map, $i: W \to X$, $\mathscr{A} \| W - \mathscr{A}$ measurable (see proposition 1, part (ii), and note that $i^{-1}[A] = A \cap W$ for $A \subset X$). An example: $\mathscr{B}(\mathbb{R})$ is the relativization of $\mathscr{B}(\mathbb{R}^*)$ and the identity $i: \mathbb{R} \to \mathbb{R}^*$ is $\mathscr{B}(\mathbb{R}) - \mathscr{B}(\mathbb{R}^*)$ measurable.

The standard method of constructing a σ-field for $X \times Y$ from σ-fields $\mathscr{A}$ for X and $\mathscr{B}$ for Y, is based on making the two projections measurable. Suppose $P_1(x, y) = x$ and $P_2(x, y) = y$ for all (x, y) in $X \times Y$. A σ-field $\mathscr{C}$ for $X \times Y$ makes P_1 a $\mathscr{C} - \mathscr{A}$ measurable function if $\mathscr{C} \supset \{P_1^{-1}[A] : A \in \mathscr{A}\} = \{A \times Y : A \in \mathscr{A}\}$ and similarly, P_2 is $\mathscr{C} - \mathscr{B}$ measurable if $\mathscr{C} \supset \{X \times B : B \in \mathscr{B}\}$. Consequently the smallest σ-field that makes both projections measurable is generated by $\{A \times Y : A \in \mathscr{A}\} \cup \{X \times B : B \in \mathscr{B}\}$, or by all "rectangles" $A \times B$ with A in $\mathscr{A}$ and B in $\mathscr{B}$. The **product borel space** of $(X, \mathscr{A})$ and $(Y, \mathscr{B})$ is

$(X \times Y, \mathscr{A} \otimes \mathscr{B})$ where $\mathscr{A} \otimes \mathscr{B}$ is the σ-field generated by the family $\{A \times B : A \in \mathscr{A}$ *and* $B \in \mathscr{B}\}$ of "rectangles". (Notice that $\mathscr{A} \otimes \mathscr{B}$ is also the δ-ring generated by the family of rectangles.)

Here are the properties of $\mathscr{A} \otimes \mathscr{B}$ that we will need.

5 PROPOSITION *If $(X, \mathscr{A})$ and $(Y, \mathscr{B})$ are borel spaces, then*

(i) *the product σ-field $\mathscr{A} \otimes \mathscr{B}$ is the smallest that makes the projections $P_1: X \times Y \to X$ and $P_2: X \times Y \to Y$ measurable,*
(ii) *$\mathscr{B}(\mathbb{R}) \otimes \mathscr{B}(\mathbb{R})$ is the Borel σ-field $\mathscr{B}(\mathbb{R}^2)$, and*
(iii) *if $(Z, \mathscr{C})$ is a borel space and $f: Z \to X \times Y$, then f is $\mathscr{C} - \mathscr{A} \otimes \mathscr{B}$ measurable iff $P_1 \circ f$ and $P_2 \circ f$ are respectively, $\mathscr{C} - \mathscr{A}$ and $\mathscr{C} - \mathscr{B}$ measurable.*

PROOF Part (i) has already been established.

The projections $P_1: \mathbb{R} \times \mathbb{R} \to \mathbb{R}$ and $P_2: \mathbb{R} \times \mathbb{R} \to \mathbb{R}$ are continuous and hence $\mathscr{B}(\mathbb{R}^2) - \mathscr{B}(\mathbb{R})$ measurable, so by part (i), $\mathscr{B}(\mathbb{R}^2) \supset \mathscr{B}(\mathbb{R}) \otimes \mathscr{B}(\mathbb{R})$. On the other hand, $\mathscr{B}(\mathbb{R}^2)$ is generated not only by the family of open sets, but also by the family of rectangles $A \times B$ with A and B open. But such sets $A \times B$ belong to $\mathscr{B}(\mathbb{R}) \otimes \mathscr{B}(\mathbb{R})$ so $\mathscr{B}(\mathbb{R}^2) \subset \mathscr{B}(\mathbb{R}) \otimes \mathscr{B}(\mathbb{R})$, and part (ii) follows.

We know that $\mathscr{A} \otimes \mathscr{B}$ is generated by $\{A \times Y : A \in \mathscr{A}\} \cup \{X \times B : B \in \mathscr{B}\}$, so $f: Z \to X \times Y$ is measurable iff $f^{-1}[A \times Y] \in \mathscr{C}$ and $f^{-1}[X \times B] \in \mathscr{C}$ for A in $\mathscr{A}$ and B in $\mathscr{B}$. But $f^{-1}[A \times Y] = (P_1 \circ f)^{-1}[A]$ and $f^{-1}[X \times B] = (P_2 \circ f)^{-1}[B]$. Consequently f is $\mathscr{C} - \mathscr{A} \otimes \mathscr{B}$ measurable iff $P_1 \circ f$ is $\mathscr{C} - \mathscr{A}$ measureable and $P_2 \circ f$ is $\mathscr{C} - \mathscr{B}$ measurable. This is the assertion of part (iii). ■

For the remainder of the section, we shall be concerned primarily with real* valued functions and we agree for convenience that a function f on X is $\mathscr{A}$ **measurable** iff f is $\mathbb{R}^*$ valued on X, $\mathscr{A}$ is a σ-field for X and f is $\mathscr{A} - \mathscr{B}(\mathbb{R}^*)$ measurable.

We deduce from the measurability of one or two functions on X, the measurability of many others by using the composition property (proposition 3, part (iii)).

6 THEOREM *If h and k are $\mathscr{A}$ measurable real valued functions on X, then each of the following is $\mathscr{A}$ measurable: any constant function, $h + k$, hk, $h \vee k$, $h \wedge k$, $h \wedge 1$, $1/h$ if h is non-vanishing, and $|h|^p$ for each positive number p.*

PROOF Let $[h, k](x) = (h(x), k(x))$ for all x in X. Then $[h, k]$ is $\mathscr{A} - \mathscr{B}(\mathbb{R}^2)$ measurable (proposition 5 part (iii)). For (a, b) in $\mathbb{R}^2$, let $+(a, b) = a + b$, $T(a, b) = ab$, $\bigvee(a, b) = a \vee b$ and $\bigwedge(a, b) = a \wedge b$. Each of the four functions $+$, T, $\bigvee$ and $\bigwedge$ is continuous on $\mathbb{R}^2$ to $\mathbb{R}^1$, hence $\mathscr{B}(\mathbb{R}^2) - \mathscr{B}(\mathbb{R})$ measurable, and so we infer by composition that

$+ \circ [h, k] = h + k$, $T \circ [h, k] = hk$, $\bigvee \circ [h, k] = h \vee k$, $\bigwedge \circ [h, k] = h \wedge k$ and $\bigwedge \circ [h, 1] = h \wedge 1$ are $\mathscr{A}$ measurable. Finally, $r \mapsto |r|^p$ is continuous and hence $\mathscr{B}(\mathbb{R}) - \mathscr{B}(\mathbb{R})$ measurable, and if $i(r) = 1|r$ for $r \neq 0$ and $i(0) = 1$ then a direct verification shows that i is $\mathscr{B}(\mathbb{R}) - \mathscr{B}(\mathbb{R})$ measurable. If follows that $|h|^p$ and, for h non-vanishing, $1/h$, are $\mathscr{A}$ measurable. ■

Each assertion of the preceding theorem remains correct if "measurable" is replaced everywhere by "continuous". No assertion of the following theorem has that property. We recall that if $\{f_n\}_n$ is a sequence of real* valued functions then $inf_n f_n$ is the function whose value at any point x is $inf_n\{f_n(x)\}$. That is, $(inf_n f_n)(x) = inf_n f_n(x)$. Similarly $(sup_n f_n)(x) = sup_n f_n(x)$, $(lim\,inf_n f_n)(x) = lim\,inf_n f_n(x)$ and $(lim\,sup_n f_n)(x) = lim\,sup_n f_n(x)$.

7 THEOREM *If $\{f_n\}_n$ is a sequence of real* valued $\mathscr{A}$ measurable functions then* $\inf_n f_n$, $\sup_n f_n$, $\liminf_n f_n$ *and* $\limsup_n f_n$ *are all $\mathscr{A}$ measurable.*

In particular, if $\{f_n\}_n$ converges pointwise to f then f is $\mathscr{A}$ measurable.

PROOF We first show that for each real number a the set $\{x: inf_n f_n(x) < a\}$ is measurable. This is true because $\{x: inf_n f_n(x) < a\} = \{x: f_n(x) < a \textit{ for some non-negative integer } n\}$ and this is the union of the sets $\{x: f_n(x) < a\}$ as n runs over the non-negative integers. Each of these sets is measurable because f_n is measurable for each n, and hence their union is measurable. Consequently $inf_n f_n$ is a measurable function.

Since $sup_n f_n = -inf_n(-f_n)$, $sup_n f_n$ is also measurable. We can see that $lim\,inf_n f_n$ is measurable by noticing that $lim\,inf_n f_n = sup_n(inf_{p>n} f_p)$ and for each n the function $inf_{p>n} f_p$ is measurable. Finally, $lim\,sup_n f_n = -lim\,inf_n(-f_n)$. ■

It will be convenient to represent an $\mathscr{A}$ measurable function as a countable linear combination of characteristic functions of members of $\mathscr{A}$. A real* valued function f on X is **$\mathscr{A}$ σ-simple** iff $\mathscr{A}$ is a family of subsets of X and $f = \sum_n a_n \chi_{A_n}$ for some sequences $\{a_n\}_n$ in $\mathbb{R}$ and $\{A_n\}_n$ in $\mathscr{A}$ (explicitly, $\{a_n \chi_{A_n}(x)\}_n$ is supposed to be summable* to $f(x)$ for each x in X). The function f is **$\mathscr{A}$ σ^+-simple** iff the sequence of coefficients can be taken to be non-negative. Every function f is $\mathscr{A}$ σ-simple for some family $\mathscr{A}$, as we now show.

For each t in $[0:\infty]$ let $D(t)$ be the "ones digit" in the (terminating if possible) dyadic representation of t and let $D(t) = 0$ for t in $[-\infty:0]$. Then $D = \sum_{n \in \mathbb{N}} \chi_{[2n-1:2n)} + \chi_{\{\infty\}}$. If $t \in [0:\infty]$, then the k-th digit in the dyadic representation of t is $D(2^{-k}t)$ and $t = \sum_{k \in \mathbb{Z}} 2^k D(2^{-k}t)$. It follows that if $f: X \to \mathbb{R}^*$, then $f^+(x) = \sum_n c_n \chi_{F_n}(f(x))$ for suitable sequences $\{c_n\}_n$ and $\{F_n\}_n$, where each c_n is positive and each F_n is of the form $[a:b)$ with $0 < a < b$, or $F_n = \{\infty\}$. Lastly we notice that $\chi_F(f(x)) = \chi_{f^{-1}[F]}(x)$.

For each function f on X to $\mathbb{R}^*$ let $\mathscr{F}_f$ consist of $\{f^{-1}[\infty]\}$ and all sets $f^{-1}[a:b)$ with $0 < a < b$, and let $\mathscr{G}_f$ consist of $\{f^{-1}[\infty]\}$ and sets $f^{-1}(a:b]$, $0 < a < b$. The preceding paragraph shows that f^+ is $\mathscr{F}_f$ σ^+-simple, and a similar argument (but use non-terminating dyadic expansions) shows f^+ is $\mathscr{G}_f$ σ^+-simple. Thus:

8 REPRESENTATION LEMMA *If f is a real* valued function, then f^+ is both $\mathscr{F}_f$ and $\mathscr{G}_f$ σ^+-simple, and f is both $\mathscr{F}_f$ and $\mathscr{G}_f$ σ-simple.*

Consequently, if $\mathscr{A}$ is a σ-field, then a real valued function f is $\mathscr{A}$ σ-simple iff it is $\mathscr{A}$ measurable.*

There is also a measure theoretic characterization of the $\mathscr{A}$ σ-simple functions when $\mathscr{A}$ is a δ-ring. A set B is **locally $\mathscr{A}$ measurable** iff $A \cap B \in \mathscr{A}$ for each member A of $\mathscr{A}$. It is straightforward to verify that the family $\mathscr{L}\mathscr{A}(X)$ (or simply $\mathscr{L}\mathscr{A}$) of locally $\mathscr{A}$ measurable subsets of X is a σ-field for X. A real* valued function f on X is **locally $\mathscr{A}$ measurable** iff it is $\mathscr{L}\mathscr{A}$ measurable. Each $\mathscr{A}$ σ-simple function is $\mathscr{L}\mathscr{A}$ σ-simple and therefore $\mathscr{L}\mathscr{A}$ measurable, but in addition: each such function vanishes outside the union of some countable subfamily of $\mathscr{A}$.

A set S is a **support** of a function f iff f vanishes outside S. We recall that for each family $\mathscr{A}$ of sets, $\mathscr{A}_\sigma$ is the family of all unions of countably many members of $\mathscr{A}$ and that if $\mathscr{A}$ is a δ-ring then $\mathscr{A}_\sigma$ is the σ-ring generated by $\mathscr{A}$, according to proposition 1.

9 THEOREM ON $\mathscr{A}$ σ-SIMPLICITY *If $\mathscr{A}$ is a δ-ring of subsets of X, then a real* valued function f is $\mathscr{A}$ σ-simple iff it is $\mathscr{L}\mathscr{A}$ measurable and has a support in $\mathscr{A}_\sigma$, or iff $f^{-1}[B] \in \mathscr{A}_\sigma$ for each B in $\mathscr{B}(\mathbb{R}^*)$ with $0 \notin B$.*

If f is non-negative and $\mathscr{A}$ σ-simple, then it is $\mathscr{A}$ σ^+-simple.

PROOF Suppose f is $\mathscr{A}$ σ-simple, say $f = \sum_n a_n \chi_{A_n}$. Then each χ_{A_n} is $\mathscr{L}\mathscr{A}$ measurable, so f is the pointwise limit of a sequence of $\mathscr{L}\mathscr{A}$ measurable functions and is therefore $\mathscr{L}\mathscr{A}$ measurable. Evidently $\bigcup_n A_n \in \mathscr{A}_\sigma$ and is a support for f.

On the other hand, suppose f is $\mathscr{L}\mathscr{A}$ measurable and has a support $\bigcup_m A_m$ with each A_m a member of $\mathscr{A}$. We may suppose $\{A_m\}_m$ disjoint. According to lemma 8, $f = \sum_n b_n \chi_{B_n}$ for some sequence $\{B_n\}_n$ in $\mathscr{L}\mathscr{A}$ and some $\{b_n\}_n$, with $b_n > 0$ if $f \geqq 0$. Then $A_m \cap B_n \in \mathscr{A}$ for all m and n because $A_m \in \mathscr{A}$ and $B_n \in \mathscr{L}\mathscr{A}$, and $f = \sum_{m,n} b_n \chi_{A_m \cap B_n}$ is required representation of f.

Suppose f is $\mathscr{L}\mathscr{A}$ measurable with support $\bigcup_n A_n$ for some $\{A_n\}_n$ in $\mathscr{A}$, and that $0 \notin B \in \mathscr{B}(\mathbb{R}^*)$. Then $f^{-1}[B] \in \mathscr{L}\mathscr{A}$ and, since $0 \notin B$, $f^{-1}[B] \subset \bigcup_n A_n$. Thus $f^{-1}[B] = \bigcup_n (A_n \cap f^{-1}[B])$, $A_n \cap f^{-1}[B]$ is in $\mathscr{A}$ for each n, and so $f^{-1}[B] \in \mathscr{A}_\sigma$.

If $f^{-1}[B] \in \mathscr{A}_\sigma$ for each B in $\mathscr{B}(\mathbb{R}^*)$ with $0 \notin B$, then $f^{-1}[B] \in \mathscr{A}_\sigma$ for sets B of each of the forms: $[a:b)$ or $(-b:-a]$ with $0 < a < b \leqq \infty$, $\{\infty\}$ or $\{-\infty\}$. Hence, by the representation lemma 8, f is $\mathscr{A}_\sigma$ σ-simple or, if $f \geqq 0$, $\mathscr{A}_\sigma$ σ^+-simple. But each member of $\mathscr{A}_\sigma$ is the union of a disjoint sequence in $\mathscr{A}$, and so f is $\mathscr{A}$ σ-simple and, if $f \geqq 0$, $\mathscr{A}$ σ^+-simple. ■

We saw in chapter 4 that each integral induces a measure. We will show, after a lemma, that each pre-integral I induces pre-measure that completely determines the minimal integral extension of I.

10 LEMMA *If $f: X \to \mathbb{R}^*$ and $a \in \mathbb{R}$, then $\{n(f \wedge a - f \wedge (a - n^{-1}))\}_n$ is a decreasing sequence that converges pointwise to $\chi_{\{x: f(x) \geqq a\}}$, and $\{n(f \wedge (a + n^{-1}) - f \wedge a)\}_n$ is increasing and converges to $\chi_{\{x: f(x) > a\}}$.*

PROOF For each x, $n(f(x) \wedge a - f(x) \wedge (a - n^{-1}))$ is 0 for $f(x) \leqq a - n^{-1}$, 1 for $f(x) \geqq a$, and is $n(f(x) - (a - n^{-1})) = 1 - n(a - f(x))$ for $a - n^{-1} \leqq f(x) \leqq a$. It follows that $\{n(f \wedge a - f \wedge (a - n^{-1}))\}_n$ is a decreasing sequence converging pointwise to $\chi_{\{x: f(x) \geqq a\}}$. A similar calculation shows that $\{n(f \wedge (a + n^{-1}) - f \wedge a)\}_n$ is increasing and converges pointwise to $\chi_{\{x: f(x) > a\}}$. ■

Suppose that J is an integral on M that extends a pre-integral I on L, that $f \in L$ and $A = \{x: f(x) \geqq 1\}$. Let $\mu_0(A) = \lim_n I(n(f \wedge 1 - f \wedge (1 - n^{-1}))) = \lim_n J(n(f \wedge 1 - f \wedge (1 - n^{-1})))$. Since J is an integral on M, lemma 10 implies that $\chi_A \in M$ and the limit is $J(\chi_A)$, so $\mu_0(A) = J(\chi_A)$ for all A in the family $\mathscr{F}$ of sets of the form $\{x: f(x) \geqq 1\}$ for $f \in L$. Thus μ_0 agrees with the measure induced by J for all A in the lattice $\mathscr{F}$, and so μ_0 is a pre-measure, the **pre-measure induced by the pre-integral *I***. Then μ_0 induces a unique measure μ on the **truncation** δ-ring of L, the δ-ring $\mathscr{T}$ generated by $\mathscr{F}$, and μ is called: the **measure induced** by I. This agrees with the earlier definition of measure induced by an integral because: if I is itself an integral on L, then $\mathscr{F} = \{A: \chi_A \in L\}$ which is a δ-ring, so the truncation δ-ring $\mathscr{T}$ coincides with $\mathscr{F}$ and $\mu(A) = \mu_0(A) = I(\chi_A)$ for A in $\mathscr{T}$. Each (non-negative) member of L is $\mathscr{T}$ σ-simple ($\mathscr{T}$ σ^+-simple, respectively) because of the representation lemma 8. If I is an integral, a pointwise sum $f = \sum_n a_n \chi_{A_n}$, with $\{a_n\}_n$ in $\mathbb{R}^+$ and $\{A_n\}_n$ in $\mathscr{T}$, belongs to L iff $\sum_n a_n \mu(A_n) < \infty$, and in this case $I(f) = \sum_n a_n \mu(A_n)$ by the Levi property.

We have proved:

11 THEOREM *If an integral J on M is an extension of a pre-integral I on L, then the measure induced by J extends the measure μ induced by I on the truncation δ-ring $\mathscr{T}$ of L.*

If I is itself an integral, then $f \in L$ iff $f(x) = \sum_n a_n \chi_{A_n}(x)$ for all x, for some $\{a_n\}_n$ in $\mathbb{R}$ and $\{A_n\}_n$ in $\mathscr{T}$ such that $\sum_n |a_n| \mu(A_n) < \infty$, and in this case $I(f) = \sum_n a_n \mu(A_n)$.

The foregoing characterization of L and I suggests a direct construction of an integral from a measure. We examine this construction in detail in the next chapter.

Finally, let us suppose that I is a pre-integral on L and $\mathscr{T}$ is its truncation δ-ring, and let L_σ, the **Baire family of L**, be $\{f: f$ *is real valued and $\mathscr{T}$ σ-simple*$\}$. In view of theorem 9, L_σ is a vector function lattice with truncation and is closed under pointwise sequential convergence. It follows from the definition of integral, that if I^1 on L^1 is the Daniell extension of I on L, $L_1 = L^1 \cap L_\sigma$ and $I_1 = I^1 | L_1$, then I_1 is an integral. In view of theorem 11, every integral J on M that extends I also extends I_1. Thus

12 COROLLARY *Each pre-integral I on L has a minimal integral extension $I_1 = I^1 | L^1 \cap L_\sigma$; that is, every integral extension of I is an extension of I_1.*

The minimal extension of a pre-integral is generally not null complete. We have seen (proposition 3.7) that the smallest *null complete* integral extension of a pre-integral is its Daniell extension.

SUPPLEMENT: STANDARD BOREL SPACES

A **Polish space** is a topological space that is homeomorphic to a complete metric separable space X, and its **Borel field** $\mathscr{B}(X)$ is the σ-field generated by the family of open sets. We will show that any two uncountable Polish spaces X and Y are **Borel isomorphic** (that is, there is a one to one map F of X onto Y such that F is $\mathscr{B}(X) - \mathscr{B}(Y)$ measurable and F^{-1} is $\mathscr{B}(Y) - \mathscr{B}(X)$ measurable). A slightly stronger result is true if X has no isolated points. In this case, there is a continuous Borel isomorphism F of the space $\mathbb{N}^\infty$ of all sequences of positive integers onto X, where $\mathbb{N}$ has the discrete topology and $\mathbb{N}^\infty$ has the product topology.

We assume throughout that X is a complete metric separable space. A point x is a **condensation** point of a subset B of X iff each neighborhood of x contains uncountably many points of B, and B is **condensed** iff each member of B is a point of condensation of B. (If B is closed in X then it is condensed iff it has no isolated points, but we shall not need this fact.) The set of all condensation points of X is a condensed subset of X.

Let us agree that a set is **convenient** iff it is non-empty, closed and condensed, and that a **convenient cover** for a set B is a sequence $\{A_n\}_n$ of convenient sets such that $A_{n+1} \setminus A_n \neq \emptyset$ for all n and $B = \bigcup_n A_n$.

13 LEMMA *If B is open in a convenient subset A of X, $B \neq \emptyset$ and $e > 0$, then there is a convenient cover $\{A_n\}_n$ of B consisting of sets of diameter less than e.*

PROOF For each x in B there is an open neighborhood V of x in A of diameter less than e whose closure V^- is a subset of B, so V^- is a convenient neighborhood of x of diameter less than e. The set B can be covered by a sequence of such convenient subsets, and after discarding each member of the sequence that is covered by its predecessors, we have either a convenient cover as desired, or a finite sequence covering B, say $A_1, A_2, \dots A_n$. Since $dia\, A_n < e$, the proof of the lemma reduces to showing that $A_n \setminus \bigcup_{k<n} A_k$ has a convenient cover. In other words, we need only show that if B is an open non-void subset of a convenient set A then B has a convenient cover.

Suppose A is a convenient set, C is a closed subset and $B = A \setminus C \neq \emptyset$. If A_n is the closure of $\{x\colon dist(x, C) > 1/n\}$ then A_n is a convenient set and $B = \bigcup_n A_n$. If we omit repetitions from the sequence $A_1, A_2, \dots A_n, \dots$ then the result is either a convenient cover for B or a finite sequence. In case of the latter, $B = A_n$ for some n, and the proof reduces to showing that a convenient set B has a convenient cover.

Suppose then that B is a convenient set, $x_0 \in B$ and that $\{x_n\}_n$ is a one to one sequence in $B \setminus \{x_0\}$ that converges to x_0. Choose a disjoint sequence $\{A_n\}_n$ such that $lim_n(\text{dia}\, A_n) = 0$ and A_n is the closure of an open neighborhood in B of x_n for each n. Then $C = \{x_0\} \cup \bigcup_n A_{2n}$ is closed, $B \setminus C$ is open, and since $x_{2n+1} \in B \setminus C$ for all n, x_0 belongs to the closure $(B \setminus C)^-$ of $B \setminus C$. Hence the sequence $(B \setminus C)^-, A_2, A_4, \dots, A_{2n}, \dots$ is a convenient cover for B. ■

14 THEOREM *If every member of a Polish space X is a point of condensation, then there is a continuous one to one map F of $\mathbb{N}^\infty$ onto X such that F^{-1} is Borel measurable.*

PROOF Let $\{A(n)\}_n$ be a convenient cover of X by sets of diameter less than one and let $B(n) = A(n) \setminus \bigcup_{k<n} A(k)$ for each n. Then $\{B(n)\}_n$ is a disjoint cover of X. Let $\{A(n, p)\}_p$ be a convenient cover of $B(n)$ by sets of diameter less than $\frac{1}{2}$ and let $B(n, p) = A(n, p) \setminus \bigcup_{q<p} A(n, q)$. Recursively, for each finite sequence $n_1, \dots, n_k$ in $\mathbb{N}$, let $\{A(n_1, \dots, n_k, p)\}_p$ be a convenient cover of $B(n_1, \dots, n_k)$ by sets of diameter less than $1/(k+1)$, and let $B(n_1, \dots, n_k, p) = A(n_1, \dots, n_k, p) \setminus \bigcup_{q<p} A(n_1, \dots, n_k, q)$.

Then for all $n_1, \dots, n_k$ and p, $B(n_1, \dots, n_k, p) \subset A(n_1, \dots, n_k, p) \subset B(n_1, \dots, n_k) \subset A(n_1, \dots, n_k)$, $\{B(n_1, \dots, n_k, q)\}_q$ is a disjoint cover of $B(n_1, \ldots, n_k)$, and $A(n_1, \ldots, n_k)$ has diameter less then $1/k$. Consequently, for each member $\nu = \{\nu_k\}_k$ of $\mathbb{N}^\infty$, $\bigcap_k B(\nu_1, \ldots, \nu_k) = \bigcap_k A(\nu_1, \ldots, \nu_k)$, and since $\{A(\nu_1, \ldots, \nu_k)\}_k$ is a decreasing sequence of closed sets whose diameters tend to zero, this intersection consists of a single point of X. We define $\boldsymbol{F(\nu)}$ to be this point.

If ν and η are members of $\mathbb{N}^\infty$ that agree for the first k coordinates, then $dist(F(\nu), F(\eta)) < 1/k$ because both $F(\nu)$ and $F(\eta)$ belong to $A(\nu_1, \ldots, \nu_k) = A(\eta_1, \ldots, \eta_k)$. Consequently F is continuous and so is a Borel mapping.

The family $\{B(n)\}_n$ is a disjoint cover of X, $\{B(n_1, \ldots, n_k, p)\}_p$ is disjoint cover of $B(n_1, \ldots, n_k)$ for all $n_1, \ldots, n_k$, and hence for each q the family of all sets $B(n_1, \ldots, n_q)$ is a disjoint cover of X. A member x of X is then $F(\nu)$ where ν is the unique sequence such that $x \in B(\nu_1, \ldots, \nu_q)$ for each q, and so F is a one to one map of $\mathbb{N}^\infty$ onto X.

Finally, the set $V(n_1, \ldots, n_k) = \{\nu\colon \nu_i = n_i \text{ for } i = 1, \ldots, k\}$ is open in $\mathbb{N}^\infty$, the family of such sets is a base for the topology of $\mathbb{N}^\infty$ and $F[V(n_1, \ldots, n_k)]$ is the Borel set $B(n_1, \ldots, n_k)$. Hence F^{-1} is Borel measurable. ■

If Z is a closed subset of a complete metric space Y and $f(y) = 1/dist(y, Z)$ for y in $Y \setminus Z$, then the graph of f, which is homeomorphic to $Y \setminus Z$, is a closed subset of $Y \times \mathbb{R}$ and is hence complete. If $\{Z_n\}_n$ is a sequence of closed subsets of Y and $(f(y))_n = 1/dist(y, Z_n)$ for each n and each y in $Y \setminus \bigcup_n Z_n$, then the graph of $f\colon Y \setminus \bigcup_n Z_n \to \mathbb{R}^\infty$ is a closed and hence complete subset of $Y \times \mathbb{R}^\infty$. In particular, if D is a countable subset of a condensed Polish space X, then $X \setminus D$ is Polish and condensed, hence Borel isomorphic to $\mathbb{N}^\infty$, and consequently X is Borel isomorphic to $X \setminus D$.

Finally, suppose that Y is an uncountable Polish space, X is the set of its points of condensation, E is the countable set $Y \setminus X$, f is a one to one map of E onto a subset D of X and g is a Borel isomorphism of X onto $X \setminus D$. Then the map that agrees with f on E and with g on X is a Borel isomorphism of Y onto X. Theorem 14 then establishes the following.

15 THEOREM *Every uncountable Polish space is Borel isomorphic to* $\mathbb{N}^\infty$.

A borel space $(X, \mathscr{A})$ that is Borel isomorphic to $(\mathbb{N}^\infty, \mathscr{B}(\mathbb{N}^\infty))$ is called a **standard borel space**.

16 NOTES See Kuratowski [1] and Engelking [1], for example, for further information in this direction. The construction given for the proof of theorem 14 is a variant of a method due to Souslin.

For a lucid, well organized account of some of the most important results on standard borel spaces see W. Arvesen, *An Invitation to C*-Algebras*, Springer-Verlag, New York, 1976.

Chapter 6

THE INTEGRAL I_μ ON $L_1(\mu)$

This section is devoted to the construction of an integral I_μ from a measure μ, to the relationships between μ and I_μ (especially for Borel measures μ for $\mathbb{R}$), and to a brief consideration of the vector spaces $L_p(\mu)$, $1 \leqq p \leqq \infty$, associated with μ.

There is no difficulty in obtaining an integral from a measure. If μ is a measure on a δ-ring $\mathscr{A}$ of subsets of X, $L^{\mathscr{A}}$ is the class of $\mathscr{A}$ simple functions on X and I^μ is the linear functional on $L^{\mathscr{A}}$ such that $I^\mu(\chi_A) = \mu(A)$ for all A in $\mathscr{A}$, then μ is also a pre-measure and therefore, according to theorem 2.7, I^μ is a pre-integral. Hence, by theorem 3.4, the Daniell extension of I^μ is an integral. Thus I^μ has an integral extension, and I_μ, the integral w.r.t. μ, is to be the minimal extension of I^μ.

The construction just outlined for I_μ is not efficient—it fails to use the fact that μ is supposed to be a measure, not just a pre-measure. We give a direct construction for I_μ and a simple proof that I_μ is an integral.

A real valued funcion f is **μ integrable** and $f \in$ **$L_1(\mu)$** (or **$L_1(X, \mathscr{A}, \mu)$** or $L_1(X, \mu)$) iff there are sequences $\{A_n\}_n$ in $\mathscr{A}$ and $\{a_n\}_n$ in $\mathbb{R}$ such that $\sum_n |a_n| \mu(A_n) < \infty$ and $f(x) = \sum_n a_n \chi_{A_n}(x)$ for every x in X, and in this case, $I_\mu(f) = \sum_n a_n \mu(A_n)$. The definition of $I_\mu(f)$ is not ambiguous, for the following reasons. The fundamental lemma 3.2 implies that if $f = \sum_n a_n \chi_{A_n} \geqq 0$ and $\sum_n I^\mu(|a_n \chi_{A_n}|) = \sum_n |a_n| \mu(A_n) < \infty$, then $\sum_n I^\mu(a_n \chi_{A_n}) = \sum_n a_n \mu(A_n) \geqq 0$. Consequently, if $f = \sum_n b_n \chi_{B_n} = \sum_n c_n \chi_{C_n}$, $\sum_n |b_n| \mu(B_n) < \infty$, and $\sum_n |c_n| \mu(C_n) < \infty$, then $\sum_n b_n \mu(B_n) - \sum_n c_n \mu(C_n) = 0$. The integral **$I_\mu(f)$** of a μ integrable function f is the **integral of f w.r.t. μ, $\int f\, d\mu$**, or **$\int f(x)\, dx$**.

The definition of $L_1(\mu)$ and I_μ can be rephrased in terms of the sequence of partial sums $s_n = \sum_{k=1}^n a_k \chi_{A_k}$ in $L^{\mathscr{A}}$ which is swiftly convergent in the sense that $\sum_n \|s_{n+1} - s_n\|_1 = \sum_n I^\mu |s_{n+1} - s_n|$ is finite. A real valued function f is μ integrable iff it is the pointwise limit of a swiftly convergent sequence $\{s_n\}_n$ in $L^{\mathscr{A}}$, and in this case $I_\mu(f)$ is $lim_n I^\mu(s_n)$.

There is another useful description of μ integrability of a non-negative function f. If f is μ integrable, then $f = \sum_n a_n \chi_{A_n}$ with $\{a_n\}_n$ in $\mathbb{R}$, $\{A_n\}_n$ in $\mathscr{A}$ and $\sum_n |a_n| \mu(A_n) < \infty$, whence $\sum_n a_n^+ \chi_{A_n}$ and $\sum_n a_n^- \chi_{A_n}$ are both μ integrable. If $f \geqq 0$ and integrable, then according to theorem 5.9, $f = \sum_n b_n \chi_{B_n}$ for some $\{b_n\}_n$ in $\mathbb{R}^+$ and $\{B_n\}_n$ in $\mathscr{A}$, and since $\sum_n a_n^+ \chi_{A_n} - \sum_{n=1}^N b_n \chi_{B_n} \geqq 0$ for each N, $\infty > \sum_n a_n^+ \mu(A_n) - \sum_{n=1}^N b_n \mu(B_n)$. Consequently $\infty > \sum_n b_n \mu(B_n) = I_\mu(f)$. We infer: a function f is non-negative and μ integrable iff there are sequences $\{b_n\}_n$ in $\mathbb{R}^+$ and $\{B_n\}_n$ in $\mathscr{A}$ so that $f = \sum_n b_n \chi_{B_n}$ and $\sum_n b_n \mu(B_n) < \infty$, and in this case $I_\mu(f) = \sum_n b_n \mu(B_n)$.

The preceding description can be rephrased in terms of sequences: a function f is non-negative and μ integrable iff it is the pointwise limit of an increasing sequence $\{s_n\}_n$ of non-negative $\mathscr{A}$ simple functions such that $lim_n I^\mu(s_n) < \infty$, and in this case $I_\mu(f) = lim_n I^\mu(s_n)$.

We use any of the foregoing descriptions of μ integrable functions as convenience dictates.

1 THEOREM (MEASURE TO INTEGRAL) *If μ is a measure on $\mathscr{A}$, then I_μ is an integral.*

Moreover, if J is an integral on M such that $\chi_A \in M$ and $J(\chi_A) = \mu(A)$ for all A in $\mathscr{A}$, then J is an extension of I_μ.

PROOF Evidently I_μ is a positive linear functional on the vector space $L_1(\mu)$. If f is the pointwise limit of a swiftly convergent sequence $\{s_n\}_n$ of $\mathscr{A}$ simple functions, then $|f| = lim_n |s_n|$ and $1 \wedge f = lim_n 1 \wedge s_n$ are also such limits, and hence $L_1(\mu)$ is a lattice with truncation.

Suppose that $\{f_n\}_n$ is a sequence of non-negative members of $L_1(\mu)$ such that $\sum_n \|f_n\| = \sum_n I_\mu(f_n) < \infty$ and $\sum_n f_n(x) < \infty$ for each x. Then for each n there are sequences $\{a_{n,k}\}_k$ in $\mathbb{R}^+$ and $\{A_{n,k}\}_k$ in $\mathscr{A}$ such that $f_n(x) = \sum_k a_{n,k} \chi_{A_{n,k}}(x)$ for all x and $I_\mu(f_n) = \sum_k a_{n,k} \mu(A_{n,k})$, whence $\sum_n f_n(x) = \sum_{n,k} a_{n,k} \chi_{A_{n,k}}(x)$ for all x and $\infty > \sum_{n,k} a_{n,k} \mu(A_{n,k}) = \sum_n I_\mu(f_n)$. It follows that $\sum_n f_n$ is μ integrable and its μ integral is $\sum_n I_\mu(f_n)$. This establishes Levi's property, and we conclude that I_μ is an integral.

Suppose J is an integral on M, $\chi_A \in M$ for all A in $\mathscr{A}$ and $J(\chi_A) = \mu(A)$. Evidently J and I_μ agree on the class $L^{\mathscr{A}}$ of $\mathscr{A}$ simple functions. Every non-negative member f of $L_1(\mu)$ is the pointwise limit of an increasing sequence $\{s_n\}_n$ in $L^{\mathscr{A}}$ and $\infty > I_\mu(f) = lim_n I(s_n) = lim_n J(s_n)$. Since J is an integral, $f \in M$ and $J(f) = I_\mu(f)$. ■

The integral with respect to μ, I_μ, is the minimal integral extension of the pre-integral I^μ, because every integral that extends I^μ is an extension of I_μ.

We notice that if $\chi_B \in L_1(\mu)$, then $\chi_B = \sum_n a_n \chi_{A_n}$ with $a_n > 0$ and A_n in $\mathscr{A}$ for each n, whence $B = \bigcup_n A_n$. Consequently B belongs to the domain of the standardization of μ. On the other hand each member C in the domain of the standardization of μ is the union $C = \bigcup_n C_n$ for a disjoint sequence $\{C_n\}_n$ in $\mathscr{A}$ with $\sum_n \mu(C_n) < \infty$, so $\chi_C \in L_1(\mu)$. It follows that the measure induced by I_μ is the standardization of the measure μ. Thus we recover the measure μ from the integral I_μ, up to standardization (see p. 48).

A further question arises naturally: Does every integral J on M occur as the integral with respect to some measure? More specifically, is $J = I_\nu$ if ν is the measure induced by J? Theorem 5.11, together with the definition of I_ν, show that this is the case. Thus:

2 THEOREM *Each integral J on M is the integral with respect to a measure, the measure induced by J.*

If μ is a measure, the measure induced by I_μ is the standardization of μ.

The preceding theorem implies that two measures yield the same integral iff they have the same standardization, and in particular, the integral with respect to a measure is identical with that with respect to its standardization.

It will be convenient to have a description of I_μ null sets in terms of μ. We recall that a set is I_μ null iff it is a subset of $E = \{x: \sum_n |f_n(x)| = \infty\}$ for some sequence $\{f_n\}_n$ in $L_1(\mu)$ with summable norms. In this case, according to lemma 3.5, $\chi_E \in L_1(\mu)$ and so by theorem 2, E belongs to the domain of the standardization of μ. Hence an I_μ null set is a subset of the union of countably many sets of μ measure zero, and it is straight forward to verify that each such set is I_μ null. We agree that a set is $\boldsymbol{\mu}$ **null** iff it is I_μ null, that a proposition holds $\boldsymbol{\mu}$ **almost everywhere** or $\boldsymbol{\mu}$ **a.e.** iff it holds except at the members of a μ null set, and that $f \geqq_\mu g$ iff $f \geqq g$ μ almost everywhere.

We state, for later convenience, a mild variant of a couple of the preceding results, as well as some corollaries of theorem 1 that follow directly from results of chapter 3. It is assumed that μ is a measure on a δ-ring of subsets of X.

3 COROLLARY (SUMMARY) *If a real valued function f on X is the pointwise limit of a sequence $\{f_n\}_n$ in $L_1(\mu)$ that is dominated by a member of $L_1(\mu)$, then $f \in L_1(\mu)$ and* $\lim_n \|f - f_n\|_1 = 0$.

Each swiftly convergent sequence $\{f_n\}_n$ in $L_1(\mu)$, and consequently each μ a.e. increasing sequence $\{f_n\}_n$ with $\sup_n I_\mu(f_n) < \infty$, *converges μ a.e. and in norm to a member of $L_1(\mu)$.*

The space $L_1(\mu)$ is norm complete; it is a norm completion of $L^{\mathscr{A}}$.

The space $L_1(\mu)$ is $\geqq_\mu$ order complete; in fact, if $\varnothing \neq W \subset L_1(\mu)$, $\sup\{I_\mu(f): f \in W\} < \infty$ *and W is closed under $\vee$, then W has a supremum.*

The null completion of $L_1(\mu)$ *is* $L_1(\mu^\vee)$, *where* $\mu^\vee$ *is the completion of* μ.

An $\mathscr{L}\mathscr{A}$ *measurable real valued function* f *is* μ *integrable if it is dominated by a member of* $L_1(\mu)$, *or if it is non negative and* $\sup_n\{I_\mu(h)$: h *is* $\mathscr{A}$ *simple and* $h \leqq f\} < \infty$.

We assume that μ is a measure on a δ-ring $\mathscr{A}$ of subsets of X and show that convergence of a sequence $\{f_n\}_n$ μ almost everywhere may imply convergence in a formally stronger sense. A sequence $\{f_n\}_n$ of real valued functions on X **converges μ almost uniformly** to a function f iff for each $e > 0$ there is a member A of $\mathscr{A}$ with $I_\mu(\chi_A) < e$ such that $\{f_n\}_n$ converges uniformly to f on $X\backslash A$. If μ is a standard measure, then $I_\mu(\chi_A) = \mu(A)$, so almost uniform convergence can be described as uniform convergence outside a set of small μ measure. In general, almost uniform convergence is uniform convergence outside the union A of a disjoint sequence $\{A_n\}_n$ in $\mathscr{A}$ with $\sum_n \mu(A_n)$ small; we sometimes say A is of **small standardized μ measure**.

Almost uniform convergence implies almost everywhere convergence; it neither implies nor is implied by convergence in L_1 norm. However, it is implied by swift convergence.

4 THEOREM (EGOROV) *If a sequence* $\{f_n\}_n$ *of* $\mathscr{A}$ σ*-simple functions converges to* $f\mu$ *a.e. and if the sequence* $\{|f_n| \wedge 1\}_n$ *is dominated* μ *a.e. by a* μ *integrable function, then* $\{f_n\}_n$ *converges to* f *almost uniformly.*

In particular, this is the case if $\{f_n\}_n$ *has a support in* $\mathscr{A}$, *or if* $\{f_n\}_n$ *is swiftly convergent.*

PROOF We assume without loss of generality that μ is a standard measure on $\mathscr{A}$ and that $f = 0$. For each n let $g_n(x) = sup\{|f_k(x)| \wedge 1$: $k \geqq n\}$. Then $\{g_n\}_n$ is a decreasing sequence of $\mathscr{A}$ σ-simple functions that converges to zero μ almost everywhere. For each $e > 0$ and for each n let $E_n = \{x\colon g_n(x) \geqq e\}$. Then $e\chi_{E_n} \leqq g_n$ and $\{g_n(x)\}_n$ is dominated by a member h of $L_1(\mu)$, so $E_n \in \mathscr{A}$ and $\mu(E_n) \leqq I_\mu(h)/e < \infty$. The sequence $\{E_n\}_n$ is decreasing and $\bigcap_n E_n$ is a null set, and consequently (since μ is continuous from above) $lim_n\, \mu(E_n) = \mu(\bigcap_n E_n) = 0$. We deduce: for $e > 0$ there is N so that $\mu(E_N) < e$, and if $n \geqq N$ and $x \notin E_N$, then $g_n(x) < e$, and hence $|f_n(x)| \wedge 1 < e$.

For each e, $0 < e < 1$, and each k, choose N_k and a set F_k in $\mathscr{A}$ such that $\mu(F_k) < e2^{-k}$ and if $n \geqq N_k$ and $x \notin F_k$, then $|f_n(x)| < e2^{-k}$. If $F = \sum_k F_k$, then $\mu(F) < e$ and for each k, if $n \geqq N_k$ and $x \notin F$, then $|f_n(x)| < e2^{-k}$. Consequently $\{f_n\}_n$ converges uniformly to 0 on $X\backslash F$. ■

We have been particularly interested in measures induced by length functions for $\mathbb{R}$. These are the Borel measures for $\mathbb{R}$; that is, measures μ on the Borel δ-ring $\mathscr{B}^\delta(\mathbb{R})$ generated by the family of compact sets (chapter 4). If μ is such a measure, then each member of $L_1(\mu)$ is $\mathscr{B}^\delta(\mathbb{R})$ σ-simple and so is locally $\mathscr{B}^\delta(\mathbb{R})$ measurable and has a support that is a countable union of members of $\mathscr{B}^\delta(\mathbb{R})$, according to theorem 5.9. But the family of locally $\mathscr{B}^\delta(\mathbb{R})$ measurable sets is just the Borel σ-field $\mathscr{B}(\mathbb{R})$, so the $\mathscr{B}^\delta(\mathbb{R})$ σ-simple functions are just the $\mathscr{B}(\mathbb{R})$ measurable real valued functions. It follows that if a $\mathscr{B}(\mathbb{R})$ measurable function f bounded by b has a bounded support, say $[-a:a]$, then f belongs to $L_1(\mu)$ because $|f| \leqq b\chi_{[-a:a]}$. Consequently the class $\mathbf{C}_c(\mathbb{R})$ of all continuous real valued functions on $\mathbb{R}$ with compact support is a subclass of $L_1(\mu)$ for each Borel measure μ for $\mathbb{R}$, and $f \mapsto I_\mu(f)$ for f in $C_c(\mathbb{R})$, is a positive linear functional on $C_c(\mathbb{R})$. Moreover:

5 RIESZ REPRESENTATION THEOREM *Each positive linear functional on $C_c(\mathbb{R})$ is the restriction to $C_c(\mathbb{R})$ of the integral with respect to a unique Borel measure μ for $\mathbb{R}$, and $C_c(\mathbb{R})$ is dense in $L_1(\mu)$.*

PROOF We know that each positive linear functional I on $C_c(\mathbb{R})$ is a pre-integral by proposition 2.9. The truncation δ-ring induced by $C_c(\mathbb{R})$ is the δ-ring generated by the family of sets $\{x: f(x) \geqq 1\}$ with f in $C_c(\mathbb{R})$ and this is just the family of compact sets. Corollary 5.12 then shows that there is one and only one Borel measure μ such that $I = I_\mu | C_c(\mathbb{R})$, and that I_μ is the minimal extension of the pre-integral I, whence $C_c(\mathbb{R})$ is dense in $L_1(\mu)$. ∎

An immediate consequence of the preceding theorem is that if μ is a Borel measure for $\mathbb{R}$ and $f \in L_1(\mu)$, then there is a continuous function g with compact support with $\|f - g\|_1$ small. Of course this also follows directly from the regularity of μ: If $E \in \mathscr{B}^\delta(\mathbb{R})$ there is by regularity a compact set K and a bounded open set U with $K \subset E \subset U$ and both $\mu(K)$ and $\mu(U)$ near $\mu(E)$. If g is a continuous function with $\chi_K \leqq g \leqq \chi_U$ (and there is such a function by Urysohn's lemma) then $\mu(K) \leqq I_\mu(g) \leqq \mu(U)$, so $\|g - \chi_E\|_1$ is small. If follows that $\mathscr{B}^\delta(\mathbb{R})$ simple functions, and hence arbitrary μ integrable functions, can be approximated in norm by members of $C_c(\mathbb{R})$.

Here is a last approximation result.

6 LUSIN'S THEOREM *If μ is a Borel measure for $\mathbb{R}$ and $f \in L_1(\mu)$, then there is a closed subset F of $\mathbb{R}$ such that $f|F$ is continuous and $\mathbb{R}\backslash F$ has small standardized μ measure.*

PROOF There is a sequence $\{f_n\}_n$ of continuous functions in $L_1(\mu)$ that converges in norm to f, and we may assume that the sequence converges swiftly. Consequently, by Egorov's theorem, there is for each $e > 0$ a set E of standardized μ measure less than $e/2$ such that $\{f_n\}_n$ converges to f uniformly on $\mathbb{R}\backslash E$. The set E is the union $\bigcup_n E_n$ of bounded Borel sets with $\sum_n \mu(E_n) < e/2$. As μ is outer regular there is for each n, a bounded open set G_n, so $G_n \supset E_n$ and $\mu(G_n \backslash E_n) < e2^{-n-1}$. Let $F = R\backslash \bigcup_n G_n$. Then $\mathbb{R}\backslash F = \bigcup_n G_n$ has standardized μ measure less than e, and $\{f_n|F\}_n$ is a sequence of continuous functions converging uniformly to $f|F$, so $f|F$ is continuous. ■

We conclude our study of the integral I_μ on $L_1(\mu)$ with a brief consideration of some other vector spaces associated with μ. The space $L_1(\mu)$ belongs to a one parameter family $L_p(\mu)$ of vector spaces constructed from μ. Suppose, for example, that μ is counting measure for $\{1, 2\}$. Then $L_1(\mu)$ can be identified with $\mathbb{R}^2$ with the norm $\|(x_1, x_2)\|_1 = |x_1| + |x_2|$. The unit ball B_1 in $L_1(\mu)$ is the "diamond" $\{(x_1, x_2): |x_1| + |x_2| \leqq 1\}$. For each (x_1, x_2) in $\mathbb{R}^2$ and for each p, $1 \leqq p < \infty$, we let $\|(x_1, x_2)\|_p = (|x_1|^p + |x_2|^p)^{1/p}$, and let $\|(x_1, x_2)\|_\infty = lim_{p\to\infty}(|x_1|^p + |x_2|^p)^{1/p} = \max\{|x_1|, |x_2|\}$. For each p, $1 \leqq p \leqq \infty$, $\| \ \|_p$ is a norm, the $L_p(\mu)$ norm for $\mathbb{R}^2$. The spaces $L_p(\mu)$ have different geometry for different values of p. The accompanying figure displays this by picturing the unit ball $\{(x_1, x_2): \|(x_1, x_2)\|_p \leqq 1\}$ for various values of p.

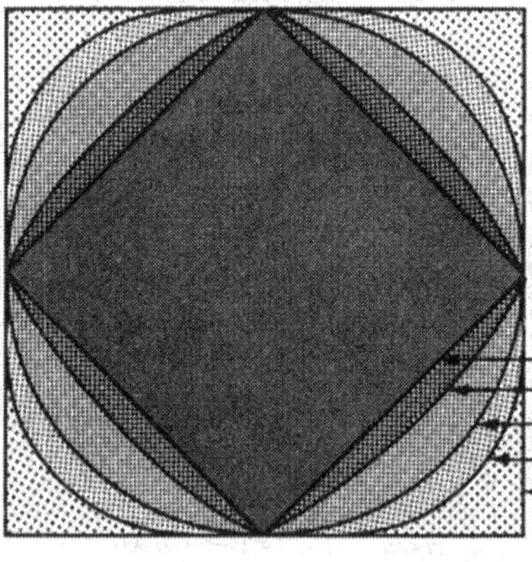

Suppose μ is an arbitrary measure on a δ-ring $\mathscr{A}$ of subsets of X and suppose for the present that $1 \leqq p < \infty$. Then $\boldsymbol{L_p(\mu)}$ is defined to be the class of real valued $\mathscr{A}$ σ-simple functions f such that $|f|^p$ is μ integrable. Thus (by 5.9) $f \in L_p(\mu)$ iff f is locally $\mathscr{A}$ measurable, has a support in $\mathscr{A}_\sigma$, and $|f|^p \in L_1(\mu)$. For such functions f, $\boldsymbol{\|f\|_p} = (\int |f|^p \, d\mu)^{1/p}$, and we call $\| \ \|_p$ the $\boldsymbol{L_p}$ **norm** (we'll prove that it's a semi-norm).

If μ is counting measure for $\{1, 2\}$, then the underlying space $L_p(\mu)$ is $\mathbb{R}^2$ for all p, but the geometry of the space (determined by the norm) varies with the p. If μ is any **totally finite** measure—that is, $sup\{\mu(A):$

$A \in \mathscr{A}\} = b < \infty$—the family $\{L_p(\mu)\}_p$ is a nested decreasing family: $L_r(\mu) \subset L_p(\mu)$ if $r > p$ (proof: if $f \in L_r(\mu)$ and $A = \{x: |f(x)| \leqq 1\}$ then $\int |f|^p \, d\mu = \int |f|^p \chi_A \, d\mu + \int |f|^p \chi_{X \setminus A} \, d\mu \leqq b + \int |f|^r \chi_{X \setminus A} \, d\mu < \infty$). For an arbitrary measure μ there is no necessary inclusion relationship among the $L_p(\mu)$ spaces for different values of p.

7 PROPOSITION *Suppose p and q are positive numbers such that $1/p + 1/q = 1$.*

(i) *If x and y are non-negative, then $xy \leqq (x^p/p) + (y^q/q)$ with equality iff $x^p = y^q$.*

(ii) (HÖLDER INEQUALITY) *If $f \in L_p(\mu)$ and $g \in L_q(\mu)$, then $fg \in L_1(\mu)$ and $\|fg\|_1 \leqq \|f\|_p \|g\|_q$ with equality iff one of $|f|^p$ and $|g|^q$ is a multiple of the other, μ almost everywhere.*

(iii) (MINKOWSKI INEQUALITY) *If f and g belong to $L_p(\mu)$, then $f + g \in L_p(\mu)$ and $\|f + g\|_p \leqq \|f\|_p + \|g\|_p$ with equality iff one of f and g is a multiple of the other, μ almost everywhere.*

PROOF (i) If $y = 0$ the lemma is clear. If $y \neq 0$, then the inequality reduces, upon dividing by y^q and letting $r = 1/p$, to $(x^p y^{-q})^r \leqq r(x^p y^{-q}) + 1 - r$. But since $0 < r < 1$, the function defined by $t \mapsto t^r$ for $t \geqq 0$, lies below its tangent line at $(1, 1)$ which is $t \mapsto rt + 1 - r$ and touches this line only at $t = 1$. The lemma follows.

(ii) Part (i) shows that $|fg| \leqq (|f|^p/p) + (|g|^q/q)$, so if $F = (|f|^p/p) + (|g|^q/q) - |fg|$ then $F \geqq 0$ and $F \in L_1(\mu)$. The set $A = \{x: F(x) > 0\}$ belongs to $\mathscr{A}_\sigma$, and if A contains a member B of $\mathscr{A}$ of positive μ measure, then $\mu(\{x: F(x) > (1/n)\})$ is positive for some n, whence $I_\mu(F) > 0$. We infer that $\|fg\|_1 \leqq ((\|f\|_p)^p/p) + ((\|g\|_q)^q/q)$ with equality iff $|f|^p = |g|^q$ μ almost everywhere. If either $\|f\|_p$ or $\|g\|_q$ is zero, the Hölder inequality is trivial. Otherwise, we replace f by $f/\|f\|_p$ and g by $g/\|g\|_q$, and so obtain $\|fg\|_1/\|f\|_p \|g\|_q \leqq (1/p) + (1/q) = 1$, with equality iff $\|g\|_q |f| = \|f\|_p |g|$ μ almost everywhere.

(iii) For convenience, let $h = f + g$. Then $|h|^p = |f + g|^p \leqq [2 \sup(|f|, |g|)]^p \leqq 2^p(|f|^p + |g|^p)$, and consequently $h \in L_p(\mu)$. We observe that $|h|^p = |h|^{p-1} |f + g| \leqq |h|^{p-1} |f| + |h|^{p-1} |g|$ with equality for $g \neq 0$ iff f is a multiple of g. We prove $\|f + g\|_p \leqq \|f\|_p + \|g\|_p$ by applying Hölder's inequality to each summand in the foregoing. Since $1/p + 1/q = 1$, $(|h|^{p-1})^q = |h|^p$, hence $|h|^{p-1} \in L_q(\mu)$, and $\| |h|^{p-1} \|_q = (\|h\|_p)^{p/q}$. Consequently Hölder's inequality shows that $I_\mu(|h|^p) \leqq (\|h\|_p)^{p/q} \|f\|_p + (\|h\|_p)^{p/q} \|g\|_p$. Hence $\|h\|_p = (I_\mu(|h|^p))^{1-1/q} \leqq \|f\|_p + \|g\|_p$. It is easy to see that equality requires that if $g \neq 0$ then f is a multiple of g μ almost everywhere. ■

Each vector space $L_p(\mu)$ shares many convergence and completeness properties with $L_1(\mu)$. We list a few of these, and outline the proofs. There are no new ideas.

8 PROPOSITION *Suppose μ is a measure on $\mathscr{A}$ and $1 \leqq p < \infty$,*

(i) (DOMINATED CONVERGENCE) *If a sequence $\{f_n\}_n$ in $L_p(\mu)$ is dominated a.e. by a member g of $L_p(\mu)$ and converges pointwise a.e. to a real valued function f, then $f \in L_p(\mu)$ and $\lim_n \|f - f_n\|_p = 0$.*

(ii) (MONOTONE CONVERGENCE) *A norm bounded, a.e. increasing sequence $\{f_n\}_n$ in $L_p(\mu)$ converges a.e. and in norm to a member f of $L_p(\mu)$ and $f(x) = \sup_n f_n(x)$ for a.e. x.*

(iii) *The class $L^{\mathscr{A}}$ of $\mathscr{A}$ simple functions is norm dense in $L_p(\mu)$.*

(iv) *Each swiftly converging sequence in $L_p(\mu)$ is dominated a.e. by a member of $L_p(\mu)$ and converges pointwise a.e. and in norm to a member of $L_p(\mu)$.*

Consequently $L_p(\mu)$ is complete.

PROOF

(i) Since $|f_n - f|^p \leqq (|f_n| + |f|)^p \leqq 2^p g^p$ a.e., $f \in L_p(\mu)$. Since $lim_n |f_n - f|^p = 0$ a.e., theorem 3.11 on dominated convergence shows that $lim_n I_\mu(|f_n - f|^p) = 0$ and hence $lim_n \|f_n - f\|_p = 0$.

(ii) We may assume that $f_n \geqq 0$ a.e. for each n (replace $\{f_n\}_n$ by $\{f_n - f_1\}_n$). Then $\{f_n{}^p\}_n$ is a norm bounded, a.e. increasing sequence in $L_1(\mu)$ and converges a.e. to a non-negative member g of $L_1(\mu)$. Then $f = g^{1/p} \in L_p(\mu)$ (check that $g^{1/p}$ is $\mathscr{L}\mathscr{A} - \mathscr{B}(\mathbb{R})$ measurable and has a support in $\mathscr{A}_\sigma$) and $\{f_n\}_n$ converges a.e. to f. The preceding result on dominated convergence then completes the proof.

(iii) Each non-negative member f of $L_p(\mu)$ is pointwise sum $\sum_n a_n \chi_{A_n}$ with $\{A_n\}_n$ in $\mathscr{A}$ and $\{a_n\}_n$ in $\mathbb{R}^+$. The preceding result shows that the sequence $\{\sum_{k=1}^n a_k \chi_{A_k}\}_n$ converges in norm to f.

(iv) Suppose $\{f_n\}_n$ is a swiftly convergent sequence in $L_p(\mu)$, and let $g_n = |f_1| + \sum_{k=1}^{n-1} |f_{k+1} - f_k|$ for each n. Then $\|g_n\|_p \leqq \|f_1\|_p + \sum_k \|f_{k+1} - f_k\|_p < \infty$ and so, by part (ii) on monotone convergence, $\{g_n\}_n$ converges a.e. to a member g of $L_p(\mu)$. Consequently, from the definition of $\{g_n\}_n$, $\{f_n\}_n$ converges a.e. to some f and since $\{f_n\}_n$ is dominated by g, $\{f_n\}_n$ converges in L_p norm to a member of $L_p(\mu)$ that agrees a.e. with f. ■

It is clear from Minkowski inequality that for f_n and f in $L_p(\mu)$, $\|f_n\|_p - \|f\|_p \leqq \|f_n - f\|_p$, and consequently, convergence in norm of a sequence $\{f_n\}_n$ to f in $L_p(\mu)$ implies convergence of the sequence of norms $\{\|f_n\|_p\}_n$ to $\|f\|_p$, that is, $f \mapsto \|f\|_p$ for f in $L_p(\mu)$ is continuous. In the reverse direction:

9 THEOREM (VITALI) *Suppose $1 \leqq p < \infty$, $\{f_n\}_n$ is a sequence in $L_p(\mu)$ converging a.e. to a member f in $L_p(\mu)$, and suppose $\{\|f_n\|_p\}_n$ converges to $\|f\|_p$. Then $\{f_n\}_n$ converges in p-norm to f.*

PROOF Since $|f_n - f|^p \leqq (|f_n| + |f|)^p \leqq 2^p(|f_n|^p + |f|^p)$, the sequence $\{2^p(|f_n|^p + |f|^p) - |f_n - f|^p\}_n$ of non-negative members of $L_1(\mu)$ converges a.e. to the member $2^{p+1}|f|^p$ whence by Fatou's lemma, $2^{p+1}\int |f|^p\,d\mu \leqq \mathit{lim\,inf}_n \int (2^p(|f_n|^p + |f|^p) - |f_n - f|^p)\,d\mu = 2^{p+1}\int |f|^p\,d\mu - \mathit{lim\,sup}_n \int |f_n - f|^p\,d\mu$. So $\mathit{lim\,sup} \int |f_n - f|^p\,d\mu = 0$. ∎

(This elegant proof is due to W. P. Novinger, *Proc. Amer. Math. Soc.* **34** (1972), 627–628.)

Two members p and q of $(1:\infty)$ are called **conjugate indices** iff $1/p + 1/q = 1$. In this case, according to Hölder's inequality, if $f \in L_p(\mu)$ and $g \in L_q(\mu)$ then $fg \in L_1(\mu)$ and $I_\mu(|f_g|) \leqq \|f\|_p \|g\|_q$. We agree that ∞ is the index conjugate to 1, and the definition of $L_\infty(\mu)$ will be such that Hölder's inequality for this pair of indices is self evident.

Suppose μ is a measure on a δ-ring $\mathscr{A}$ of subsets of X. We recall that the family $\mathscr{L}\mathscr{A}$ of locally measurable sets is $\{B\colon B \cap A \in \mathscr{A}$ *for all* A *in* $\mathscr{A}\}$. A set E is **locally μ null** iff E is a subset of a member B in $\mathscr{L}\mathscr{A}$ and $\mu(B \cap A) = 0$ for all A in $\mathscr{A}$, and a property holds **locally μ a.e.** iff it holds outside some locally μ null set. A real valued function f on X is **μ essentially bounded** or just **essentially bounded** iff for some r, $|f| \leqq r$ locally μ a.e., and for such a function, $\|f\|_\infty = \mathit{inf}\{r\colon |f| \leqq r$ *locally* μ *a.e.*$\}$. We notice that $\|f\|_\infty \leqq \|f\|_X = \mathit{sup}_{x \in X}|f(x)|$ and that $\|f\|_\infty$ may be zero while $\|f\|_X = \infty$. The space $\boldsymbol{L_\infty(\mu)}$ or $\boldsymbol{L_\infty(\mu, X)}$ is the class of $\mathscr{L}\mathscr{A}$ measurable essentially bounded functions; it is entirely determined by the δ-ring $\mathscr{A}$ and the family of locally μ null sets.

10 PROPOSITION *The class $L_\infty(\mu)$ is a complete semi-normed space.*

Each member of $L_\infty(\mu)$ that is bounded on X by b in absolute value, is the uniform limit of a sequence of $\mathscr{L}\mathscr{A}$ simple functions whose absolute value is bounded by b. Consequently the family of $\mathscr{L}\mathscr{A}$ simple functions is dense in $L_\infty(\mu)$.

PROOF Evidently $L_\infty(\mu)$ is a vector space. We notice that if $f \in L_\infty(\mu)$ then $|f| \leqq \|f\|_\infty$ locally μ a.e., because for each n, $|f| \leqq \|f\|_\infty + 1/n$ outside some locally μ null set B_n, and $\bigcup_n B_n$ is locally μ null.

If $\{f_n\}_n$ is a Cauchy sequence in $L_\infty(\mu)$ then $|f_p(x) - f_q(x)| \leqq \|f_p - f_q\|_\infty$ except for x in a locally μ null set $A_{p,q}$. Let $Y = X \setminus \bigcup_{p,q} A_{p,q}$, and for each n let $g_n = f_n$ on Y and 0 on $X \setminus Y$. Then $\{g_n\}_n$ is a sequence in $L_\infty(\mu)$, $\|f_n - g_n\|_\infty = 0$ for each n, and since $|g_p(x) - g_q(x)| \leqq \|f_p - f_q\|_\infty$ for all p, q and x, $\{g_n\}_n$ is a Cauchy sequence with respect to the sup norm $\|\ \|_X$. Consequently $\{g_n\}_n$ converges in the sup norm, and hence in the $L_\infty(\mu)$ norm, to a bounded locally $\mathscr{A}$ measurable function g. Therefore $\mathit{lim}_n \|f_n - g\|_\infty = 0$.

Finally suppose f is $\mathscr{L}\mathscr{A}$ measurable, $0 \leqq f(x) \leqq b$ and $n \geqq b$. Then the function $s_n = \sum_{k=1}^{n\cdot 2^n} (k-1)2^{-n}\chi_{f^{-1}[(k-1)2^{-n}:k2^{-n}]}$ is $\mathscr{L}\mathscr{A}$ simple and $|f - s_n| \leqq 2^{-n}$. The proposition follows. ■

The spaces $L_p(\Lambda^1)$ for Lebesgue measure Λ^1 in $\mathbb{R}$ are of special interest. It is easy to see that for $1 \leqq p < \infty$, the family of $\mathscr{J}$ simple functions, where $\mathscr{J}$ is the collection of closed intervals, is dense in $L_p(\Lambda^1)$, and so is the family $C_c(\mathbb{R})$ of continuous functions with compact support. For each real valued function f on $\mathbb{R}$ and each t in $\mathbb{R}$, suppose $T_t(f)$ is the translate of f by t; that is, $T_t(f)(x) = f(x+t)$ for x in $\mathbb{R}$. If $f \in L_p(\Lambda^1)$ so does $T_t(f)$ and $\|T_t(f)\|_p = \|f\|_p$, because Λ^1 is translation invariant. Thus, for each t in $\mathbb{R}$, T_t is an isometric map of $L_p(\Lambda^1)$ onto itself.

We show that for each f in $L_p(\Lambda^1)$, $t \to T_t(f)$ for t in $\mathbb{R}$ is a uniformly continuous map of $\mathbb{R}$ to $L_p(\Lambda^1)$. This will follow if we show that $\|T_h(f) - f\|_p$ is small for $|h|$ small, since $\|T_s(f) - T_t(f)\|_p = \|T_{s-t}(f) - f\|_p$.

11 Theorem (Continuity of Translation on $L_p(\Lambda^1)$)
If $f \in L_p(\Lambda^1)$ and $1 \leqq p < \infty$, then $\lim_{|h|\to 0} \|T_h(f) - f\|_p = 0$.

PROOF Suppose $g \in C_c(\mathbb{R})$ and $[a:b]$ is a support for g. Then $[a - |h| : b + |h|]$ is a support for $|T_h(g) - g|$ so $\|T_h(g) - g\|_p \leqq ((b - a + 2|h|)\, sup_x |T_h(g)(x) - g(x)|^p)^{1/p}$, and the supremum is small for $|h|$ small because g is uniformly continuous. Thus $t \mapsto T_t(g)$ is continuous.

For $e > 0$, choose g in $C_c(\mathbb{R})$ so $\|f - g\|_p < e$, and take $|h|$ small enough that $\|T_h(g) - g\|_p < e$. Then $\|T_h(f) - f\|_p \leqq \|T_h(f) - T_h(g)\|_p + \|T_h(g) - g\|_p + \|g - f\|_p < 3e$. The theorem follows. ■

We note that the preceding theorem fails for $p = \infty$—e.g., let f be the characteristic function $\chi_{(0:\infty)}$.

We conclude with a single application of the theorem that $L_2(\mu)$ is complete. This space has a particularly interesting structure. If f and g belong to $L_2(\mu)$ then, according to Hölder's inequality, fg is μ integrable and $I_\mu(|fg|) \leqq \|f\|_2 \|g\|_2$, with equality iff $\|g\|_2 |f| = \|f\|_2 |g|$ μ almost everywhere. We let the **inner product** of f and g, $\langle f, g \rangle$, be $I_\mu(fg)$. Then $(f,g) \mapsto \langle f, g\rangle$ is linear in each variable, symmetric in the sense that $\langle f, g\rangle = \langle g, f\rangle$, and $\langle f, f\rangle \geqq 0$ for all f. Such a space is called a **real euclidean space**. Thus $L_2(\mu)$, with $\langle\ ,\ \rangle$, is a real euclidean space, $\|f\|_2 = (\langle f, f\rangle)^{1/2}$ is a semi-norm, and $L_2(\mu)$ with this semi-norm is complete.

A **complex euclidean space** is a vector space E over $\mathbb{C}$ with an inner product $\langle\ ,\ \rangle$ on $E \times E$ to $\mathbb{C}$ such that $(x, y) \mapsto \langle x, y\rangle$ is linear in x and conjugate linear in y (i.e., $\langle x, ay + bz\rangle = a^{\sim}\langle x, y\rangle + b^{\sim}\langle x, z\rangle$ for all x, y and z in E and all a and b in $\mathbb{C}$, where $^{\sim}$ denotes complex conjugation);

$\langle y, x\rangle$ is the complex conjugate of $\langle x, y\rangle$ for all x and y; and $\langle x, x\rangle \geqq 0$ for all x.

If E with $\langle\ ,\ \rangle$ is a real or complex euclidean space, then the **euclidean semi-norm** is given by $\|x\| = (\langle x, x\rangle)^{1/2}$. This definition does yield a semi-norm for the following reason. If x and y belong to E and a is a complex number of modulus 1 such that $\langle ax, y\rangle = |\langle x, y\rangle|$, then $0 \leqq \langle a\|y\|x - \|x\|y, a\|y\|x - \|x\|y\rangle = 2\|x\|^2\|y\|^2 - 2|\langle x, y\rangle|\,\|x\|\,\|y\|$, whence (the **Cauchy-Schwartz** inequality) $|\langle x, y\rangle| \leqq \|x\|\,\|y\|$ with equality holding iff $\|(a\|y\|x - \|x\|y)\| = 0$. Consequently $\|x + y\|^2 = \langle x + y, x\rangle + \langle x + y, y\rangle \leqq \|x + y\|\,\|x\| + \|x + y\|\,\|y\|$, so $\|x + y\| \leqq \|x\| + \|y\|$ with equality iff $\|y - bx\| = 0$ for some complex number b.

A linear functional F on a real or complex semi-normed space E is **bounded** iff $sup\{|F(x)|: \|x\| \leqq 1\} < \infty$, and in this case $\|F\|$ is defined to be this supremum. It is easy to verify that F is bounded iff it is continuous relative to the semi-metric $(x, y) \mapsto \|x - y\|$. If E is a euclidean space, $y \in E$ and $F(x) = \langle x, y\rangle$ for all x in E, then F is a bounded linear functional by reason of the Cauchy–Schwartz inequality, and in fact $\|F\| = \|y\|$. We will show that every bounded linear functional on a complete euclidean space is of the form $x \mapsto \langle y, x\rangle$ for some y.

We need a preliminary lemma and this lemma depends on the **parallelogram law**: if x and y belong to a euclidean space, then $\|x + y\|^2 + \|x - y\|^2 = 2\|x\|^2 + 2\|y\|^2$. This follows directly by "expanding" $\langle x + y, x + y\rangle + \langle x - y, x - y\rangle$. We agree that a member x of E is **perpendicular** to a subset H iff $\langle x, y\rangle = 0$ for all y in H.

12 PROPOSITION *If H is a closed vector subspace of a complete euclidean space E and $x_0 \in E\backslash H$, then there is y_0 in H such that $x_0 - y_0$ is perpendicular to H.*

PROOF Choose a sequence $\{y_n\}_n$ in H such that $lim_n \|x_0 - y_n\| = inf_{y \in H}\|x_0 - y\| = K$. We assert that $\{y_n\}_n$ is a Cauchy sequence. Indeed, for $e > 0$ if N is an integer such that $K + e \geqq \|x_0 - y_n\| \geqq K$ for $n \geqq N$, and if we set $z_k = x_0 - y_k$ for all k, then $\|y_N - y_n\|^2 = \|z_N - z_n\|^2 = 2\|z_N\|^2 + 2\|z_n\|^2 - \|z_N + z_n\|^2 = 2\|x_0 - y_N\|^2 + 2\|x_0 - y_n\|^2 - 4\|x_0 - (y_N + y_n)/2\|^2 \leqq 2(K + e)^2 + 2(K + e)^2 - 4K^2$, since $(y_N + y_N)/2 \in H$, so $\|y_N - y_n\|^2 < 8Ke + 4e^2$. It follows that $\{y_n\}_n$ is a Cauchy sequence. This sequence converges to a member y_0 of H because E is complete and H is closed. Thus the infimum of $\|x_0 - y\|$ for y in H is assumed at $y = y_0$.

We infer that if z_0 is the non-zero vector $x_0 - y_0$, then $\|z_0\|^2 \leqq \|z_0 - v\|^2$ for all v in H. Thus the quadratic function $t \to \langle z_0 - tv, z_0 - tv\rangle - \langle z_0, z_0\rangle$ for t in $\mathbb{R}$, has a minimum at $t = 0$, and so the coefficient of t, which is $-\langle z_0, v\rangle - \langle v, z_0\rangle$, is zero for each v in H. There is a complex number a such that $\langle z_0, av\rangle = |\langle z_0, v\rangle|$, and since $av \in H$, $\langle z_0, v\rangle = 0$ for each v in H. ■

We agree that a complete euclidean space is a **Hilbert Space**. If F is a bounded linear functional on the (real or complex) Hilbert space E, then $H = \{x\colon F(x) = 0\}$ is closed. If F is not identically zero (otherwise $F(x) = \langle x, 0\rangle$ for all x), then $H \neq E$ and so there is by proposition 12 a non-zero vector u that is perpendicular to H, and since $u \notin H$, $F(u) \neq 0$. Then for x in E, $x - (F(x)/F(u))u \in H$ so $\langle x - (F(x)/F(u))u), u\rangle = 0$, whence $\langle x, u\rangle = (F(x)/F(u))\langle u, u\rangle$ so $F(x) = (F(u)/\langle u, u\rangle)\langle x, u\rangle$ for all x. This establishes the following.

13 THEOREM (RIESZ) *Each bounded linear functional F on a Hilbert space E is of the form $F(x) = \langle x, u\rangle$, for some u.*

In particular, if F is a bounded linear functional on $L_2(\mu)$, then there is g in $L_2(\mu)$ such that $F(f) = \int fg\,d\mu$ for all f in $L_2(\mu)$.

SUPPLEMENT: BOREL MEASURES AND POSITIVE FUNCTIONALS

It is assumed throughout that **X is a locally compact Hausdorff** space. A Borel measure for X is a measure μ on the δ-ring $\mathscr{B}^\delta(X)$ generated by compact subsets of X. The measure is regular iff members of $\mathscr{B}^\delta(X)$ have inner μ approximations by compacta, and this is the case iff members have outer μ approximations by open members of $\mathscr{B}^\delta(X)$. (See the Supplement: Measures on $\mathscr{B}^\delta(X)$, to chapter 4.)

Each closed set belongs to the σ-field $\mathscr{L}\mathscr{B}^\delta(X)$ of locally $\mathscr{B}^\delta(X)$ measurable sets, and consequently each real valued continuous function on X is $\mathscr{L}\mathscr{B}^\delta(X)$ measurable. If such a function f has a compact support K, then f is dominated by a scalar multiple of χ_K and so f is μ integrable for every Borel measure μ. Thus the class $C_c(X)$ of continuous real valued functions f on X with compact support is a subclass of $L_1(\mu)$ for each Borel measure μ. If μ is regular, then $C_c(X)$ is dense in $L_1(\mu)$.

14 PROPOSITION *If μ is a regular Borel measure for X, then $C_c(X)$ is dense in $L_1(\mu)$.*

PROOF It is sufficient to show that for each B in $\mathscr{B}^\delta(X)$, there is f in $C_c(X)$ so that $\|f - \chi_B\|_1$ is small, because linear combinations of finitely many such functions are dense in $L_1(\mu)$ (by the definition of $L_1(\mu)$). There is a compact set K and an open member U of $\mathscr{B}^\delta(X)$ such that $K \subset B \subset U$ and $\mu(U) - \mu(K)$ is small, and Urysohn's lemma implies that there is f in $C_c(X)$ with $\chi_K \leqq f \leqq \chi_U$, whence $\mu(K) \leqq I_\mu(f) \leqq \mu(U)$. But we also have $\chi_K \leqq \chi_B \leqq \chi_U$ so $\mu(K) \leqq I_\mu(\chi_B) \leqq \mu(U)$, whence both $\chi_B - f$ and $f - \chi_B$ are between $-(\chi_U - \chi_K)$ and $\chi_U - \chi_K$ and $\|f - \chi_B\|_1 \leqq \mu(U) - \mu(K)$ which is small. Consequently $C_c(X)$ is dense in $L_1(\mu)$. ■

The preceding proposition makes it possible to establish a straightforward generalization of theorem 11. Suppose G is a locally compact Hausdorff topological group, η is a left invariant regular Borel measure (a left Haar measure), $1 \leqq p < \infty$ and $f \in L_p(\eta)$. For each h in G, let $T_h(f)(x) = f(hx)$ for all x in G. Then $T_h(f)$ is near $T_k(f)$ in $L_p(\eta)$ provided h is near k in G (i.e., provided $h^{-1}k$ is near the identity e). Rephrased: $h \to T_h(f)$ is a continuous map of G into $L_p(\eta)$.

15 THEOREM ON CONTINUITY OF TRANSLATION *If η is a left Haar measure for G, $1 \leqq p < \infty$ and $f \in L_p(\eta)$, then $\|T_h(f) - f\|_p$ is small for h near e.*

PROOF We first establish the theorem for a member g of $C_c(G)$. Because g has a compact support, it is uniformly continuous in the sense that for $\varepsilon > 0$ there is a neighborhood W of e such that $|g(y) - g(x)| < \varepsilon$ if $yx^{-1} \in W$. In other words, if $h \in W$ then $|T_h(g)(x) - g(x)| < \varepsilon$ for all x. Thus $T_h(g)$ converges to g uniformly as h converges to e. If K is a compact support for g and V is a compact neighborhood of e, then $V^{-1}K$ is a support for $T_h(g)$ for every h in V, and since $\eta(V^{-1}K) < \infty$, it follows that $T_h(g)$ converges to g in the norm $\| \ \|_p$.

If $f \in L_p(\eta)$ and g is a member of $C_c(G)$ such that $\|f - g\|_p$ is small, then since $\|T(f) - T(g)\|_p = \|f - g\|_p$ and $\|T_h(f) - f\|_p \leqq \|T_h(f) - T_h(g)\|_p + \|T_h(g) - g\|_p + \|g - f\|_p$, $\|T_h(f) - f\|_p$ is near $\|T_h(g) - g\|_p$. The theorem follows. ■

If μ is a Borel measure for X, then the map $f \mapsto \int f\,d\mu$ for f in $C_c(X)$, is a positive linear functional on $C_c(X)$. If $C_0(X)$ is the space of those continuous real valued functions on X that vanish at ∞, and v is a Borel measure for X, then $f \mapsto \int f\,dv$ also defines a positive linear function on $C_0(X)$, *provided* v is **totally finite**—that is, $sup\{v(A): A \in \mathscr{B}^\delta(X)\} = \|v\|_V < \infty$. (We leave the proof of this fact to the reader.) Both of these statements have converses, as does proposition 14.

16 RIESZ REPRESENTATION THEOREM *For each positive linear functional F on $C_c(X)$, there is a unique regular Borel measure μ such that $F(f) = \int f\,du$ for all f in $C_c(X)$.*

A positive linear functional J on $C_0(X)$ is also of the form $f \mapsto \int f\,dv$ for a unique regular Borel measure v and, moreover, $\|J\| = \|v\|_V$.

PROOF A positive linear functional F on $C_c(X)$ is a pre-integral according to proposition 2.10, and theorem 5.11 then asserts that there is a unique measure ρ on the truncation δ-ring $\mathscr{T}$ of $C_c(X)$, the measure induced by F, such that the integral I_ρ is an extension of F. The truncation δ-ring $\mathscr{T}$ is generated by the lattice $\mathscr{F}$ of sets of the form $\{x: f(x) \geqq 1\}$ with f in $C_c(X)$. Every compact set K has a compact neighborhood that belongs to $\mathscr{F}$, and $\rho|\mathscr{F}$ is non-negative, monotonic, additive and subadditive—it is a pre-content in the terminology of chapter 1. Moreover, since $\rho(\{x: f(x) \geqq 1\}) = \rho(\bigcap_n \{x: f(x) \geqq 1 - (1/n\})$ and $\{x: f(x) \geqq 1 - (1/n)\}$ is a compact neighborhood of $\{x: f(x) \geqq 1\}$, the regularization of $(\rho|\mathscr{F})$ is an extension of $\rho|\mathscr{F}$. Consequently, the Borel measure μ that extends the regularization of $\rho|\mathscr{F}$ is a regular Borel extension of ρ, and hence $F(f) = \int f\,d\rho = \int f\,d\mu$ for all f in $C_c(X)$. Suppose π is any regular Borel measure such that $F(f) = \int f\,d\pi$ for f in $C_c(X)$. Then by the preceding proposition, $C_c(X)$ is dense in $L_1(\mu)$ and in $L_1(\pi)$ in their norms and the norms agree on $C_c(X)$. It follows that $\pi = \mu$.

If J is a positive linear functional on $C_0(X)$, then there is certainly a unique, regular Borel measure ν so that $J(f) = \int f\,d\nu$ for f in $C_c(X)$. Because J is positive it is bounded (see chapter 2) and if f vanishes off K and $\|f\|_X \leqq 1$, $|J(f)| \leqq \mu(K) \leqq \|\nu\|_V$, so $\|J\| \leqq \|\nu\|_V$. On the other hand if $B \in \mathscr{B}^\delta(X)$, K is a compact inner approximation for B and f is a member of $C_c(X)$ which is 1 on K and $0 \leqq f \leqq 1$, then $\nu(B) \leqq \nu(K) + e < \int f\,d\nu + e = J(f) + e$ whence $sup\{\nu(B): B \in \mathscr{B}^\delta(X)\} \leqq \|J\|$, leading to equality. Finally, J and $f \mapsto \int f\,d\nu$ for f in $C_0(X)$, are both bounded linear functionals that agree on the dense subspace $C_c(X)$ of $C_0(X)$, and hence they agree on $C_0(X)$. ■

Here is a simple corollary. Suppose ν is a Borel measure such that $C_c(X)$ is dense in $L_1(\nu)$ (relative to the I_ν norm), and suppose μ is the regular Borel measure such that $\int f\,d\mu = \int f\,d\nu$ for all f in $C_c(X)$. Then $C_c(X)$ is also dense in $L_1(\mu)$ (relative to the I_μ norm) and the two norms agree on $C_c(X)$. It follows that $\nu = \mu$. Thus

17 Corollary *If ν is a Borel measure for X and $C_c(X)$ is dense in $L_1(\nu)$ then ν is regular.*

Here is one more approximation result for a function that is integrable w.r.t. a regular Borel measure μ—a generalization of theorem 6. We recall that the standardized μ measure of a set A is v iff $A = \sum_n A_n$ for some disjoint sequence $\{A_n\}_n$ in $\mathscr{B}^\delta(X)$ such that $v = \bigcup_n \mu(A_n)$—that is, iff the standardization of μ takes the value v at A. A set X is **σ-compact** iff it is the union of countably many compact sets.

18 Lusin's Theorem *If μ is a regular Borel measure for X, $f \in L_1(\mu)$ and $e > 0$, then there is g in $C_0(X)$ and an open set U such that $g = f$ on $X \setminus U$, $\|f - g\|_1 < e$ and U has standardized μ measure less than e.*

If X is σ-compact, then any $\mathscr{L}\mathscr{B}^\delta(X)$ measurable function f is continuous on $X \setminus U$ for some open set U of small standardized μ measure.

PROOF We assume (without loss of generality) that $f \geqq 0$, and for convenience in statement, we fail to distinguish between μ and its standardization (the measure induced by I_μ).

Suppose $f \in L_1(\mu)$, and for each n let $U_n = \{x: f(x) \geqq n\}$. Then $n\chi_{U_n} \leqq f$, so $\mu(U_n) \leqq \|f\|_1/n$, and $f = n \wedge f$ except at the points of a set U_n which, for n large, has small μ measure. Because μ is regular, U_n is a subset of an open set V_n of small measure. Moreover, the increasing sequence $\{n \wedge f\}_n$ converges pointwise to f and hence $\|n \wedge f - f\|_1$ is small for n large. Consequently, for $e > 0$ there is N in $\mathbb{N}$ and an open set V such that $\|f - N \wedge f\|_1 < e/2$, $\mu(V) < e/2$ and $f = N \wedge f$ on $X \setminus V$.

We show that there is an open set W such that $\mu(W) < e/2N$ and f is continuous on $X \setminus W$. Then on $X \setminus W$, $N \wedge f$ is continuous, non-negative, bounded by N, and has a σ-compact support. It follows that $N \wedge f$ has a continuous extension $g: X \to [0:N]$ by Tietze's theorem. Then $\mu(V \cup W) < e$ and $\|f - g\|_1 \leqq \|f - N \wedge f\|_1 + \|N \wedge f - g\|_1 < e/2 + N(e/2N) = e$.

To establish the existence of the desired open set W: Choose a sequence $\{f_n\}_n$ in $C_c(X)$ that converges swiftly to f, invoke Egorov's theorem 4 to see that $\{f_n\}_n$ converges to f almost uniformly, and use the fact that the uniform limit of continuous functions is continuous.

The last statement of the theorem follows from the first together with the $e/2^n$ argument. Suppose that $\{K_n\}_n$ is an increasing sequence of compact sets with $X = \bigcup_n K_n$, that K_{n+1} is a neighborhood of K_n, let $f_n = f \wedge n\chi_{K_n}$ and choose an open set V_n such that f_n is continuous on $X \setminus V_n$ and $\mu(V_n) < e/2^n$. Then $V = \bigcup_n V_n$ has measure less than e, and f is continuous on $X \setminus V$. ■

Chapter 7

INTEGRALS* AND PRODUCTS

It will be convenient to extend the domain of an integral to include certain $\mathbb{R}^*$ valued functions, and to extend the integral to an $\mathbb{R}^*$ valued functional on the larger domain. We make this extension and subsequently phrase the Beppo Levi theorem and Fatou's lemma in this context. A more serious use of the new construct is then made in the study of product integrals and product measures.

We recall that a sequence $\{t_n\}_n$ of members of $\mathbb{R}^*$ is **summable*** or **summable in the extended sense**, iff $\{\sum_{n \in F} t_n\}_F$ has a limit in $\mathbb{R}^*$ as F runs through the family of finite subsets of $\mathbb{N}$ (the family is directed by inclusion), and in this case, $\sum_n t_n$ is $lim\{\sum_{n \in F} t_n : F \subset \mathbb{N}\}$ and $\{t_n\}_n$ is summable* to $\sum_n t_n$. Thus, the sequence $\{t_n\}_n$ is summable iff it is summable* and $\sum_n t_n$ is finite, and the sequence $\{t_n\}_n$ is summable* *unless* $\sum_n (t_n \vee 0) = \infty$ and $\sum_n (t_n \wedge 0) = -\infty$.

Suppose μ is a measure on a δ-ring $\mathscr{A}$ of subsets of X and f is an $\mathbb{R}^*$ valued function. Then f is **μ integrable***, or **integrable***, or **integrable in the extended sense**, iff for some sequence $\{A_n\}_n$ in $\mathscr{A}$ and some $\{a_n\}_n$ in $\mathbb{R}$, $\{a_n \mu(A_n)\}_n$ is summable* and $\{a_n \chi_A(x)\}_n$ is summable* to $f(x)$ for each x. The class of all μ integrable* functions is denoted $\boldsymbol{L^*(\mu)}$. Evidently $L_1(\mu) \subset L^*(\mu)$.

Each μ integrable function f is real valued and satisfies a measurability condition, that f be $\mathscr{A}$ σ-simple (or equivalently, f is locally $\mathscr{A}$ measurable with an $\mathscr{A}_\sigma$ support). Each μ integrable* function is real* valued and satisfies the same measurability condition.

Each real valued $\mathscr{A}$ σ-simple function f that is dominated by a member g of $L_1(\mu)$ (that is, $|f| \leqq g$) belongs to $L_1(\mu)$. We show that each $\mathbb{R}^*$ valued $\mathscr{A}$ σ-simple function that is bounded above, or below, μ a.e. by a member of $L_1(\mu)$, itself belongs to $L^*(\mu)$.

1 Proposition *An $\mathbb{R}^*$ valued function is μ integrable* iff it is locally $\mathscr{A}$ measurable with a support in $\mathscr{A}_\sigma$ and is bounded either below or above μ a.e. by a μ integrable function.*

Consequently $L^*(\mu)$ is a function lattice with truncation, and if $\{f_n\}_n$ is a sequence in $L^*(\mu)$ that is bounded below or above μ a.e. by a member of $L_1(\mu)$, then $\sup_n f_n$, $\inf_n f_n$, $\limsup_n f_n$ and $\liminf_n f_n$ belong to $L^*(\mu)$.

PROOF Suppose that f is integrable*, $f = \sum_n a_n \chi_{A_n}$ with $\{a_n \mu(A_n)\}_n$ summable*, $g = \sum_n a_n^+ \chi_{A_n}$ and $h = \sum_n a_n^- \chi_{A_n}$. Then f is bounded above by g and below by $-h$, and g or h is equal μ a.e. to an integrable function according as the sequence $\{a_n^+ \mu(A_n)\}_n$ or $\{a_n^- \mu(A_n)\}_n$ is summable.

Conversely if f is locally $\mathscr{A}$ measurable with an $\mathscr{A}_\sigma$ support, then so are f^+ and f^- whence $f^+ = \sum_n a_n \chi_{A_n}$ and $f^- = \sum_n b_n \chi_{B_n}$ with $a_n \geqq 0$, $b_n \geqq 0$, $A_n \in \mathscr{A}$ and $B_n \in \mathscr{A}$ for each n. If f is bounded below μ a.e. by an integrable function v, then $f^- \leqq v^-$ μ a.e., v^- is integrable, so $\{b_n \mu(B_n)\}_n$ is summable. Likewise if f is bounded above μ a.e. by a member of $L_1(\mu)$ then $\{a_n \mu(A_n)\}_n$ is summable. It follows that f is integrable* in either case. ■

We notice from the foregoing that if f is integrable*, $f = \sum_n a_n \chi_{A_n}$, and f is not bounded above μ a.e. by a member of $L_1(\mu)$, then $\sum_n a_n^+ \mu(A_n) = \infty$ and hence $\sum_n a_n \mu(A_n) = \infty$. Similarly if f is not bounded below μ a.e. by an integrable function then $\sum_n a_n \mu(A_n) = -\infty$. On the other hand if f is bounded both above and below μ a.e. by integrable functions, then f is μ a.e. equal to an integrable function g, and it is easy to see that $\sum_n a_n \mu(A_n) = I_\mu(g)$.

For each member f of $L^*(\mu)$ we define the **extended integral $I_\mu^*(f)$** to be $\sum_n a_n \mu(A_n)$, provided $f(x) = \sum_n a_n \chi_{A_n}(x)$ for each x and $\{a_n \mu(A_n)\}_n$ is summable*. We have just seen that this definition is not ambiguous, that $I_\mu^*(f)$ is finite iff f is equal μ a.e. to a member of $L_1(\mu)$, and that $I_\mu^* | L_1(\mu) = I_\mu$. If $I_\mu^*(f) = \infty$ (or $-\infty$) then f^- (respectively f^+) agrees a.e. with an integrable function. We agree that if $f \in L^*(\mu)$, then $\int f\,d\mu$, or $\int f(x)\,d\mu x$, is $I_\mu^*(f)$.

If f and g are members of $L^*(\mu)$ and $f \leqq g$ μ a.e., then $I_\mu^*(f) \leqq I_\mu^*(g)$. This is the only additional fact needed to establish the following convenient form of the monotone convergence theorem for I_μ^*.

2 Theorem (*B. Levi*) *If $\{f_n\}_n$ is an increasing sequence in $L^*(\mu)$ and $I_\mu^*(f_1) > -\infty$, then $\lim_n f_n \in L^*(\mu)$ and $I_\mu^*(\lim_n f_n) = \lim_n I_\mu^*(f_n)$.*

Consequently, if $\{f_n\}_n$ is in $L^*(\mu)$ and $f_n \geqq 0$ for each n, then $\sum_n f_n \in L^*(\mu)$ and $I_\mu^*(\sum f_n) = \sum I_\mu^*(f_n)$.

PROOF Since $I_\mu^*(f_1) > -\infty$, f_1^- agrees a.e. with an integrable function, and since $f_n \geqq f_1 \geqq -f_1^-$ for each n, $\lim_n f_n \in L^*(\mu)$ by the preceding proposition. It is evident that $I_\mu{}^*(\lim_n f_n) \geqq \lim_n I_\mu{}^*(f_n)$, so equality holds if the limit on the right is ∞. Suppose that $\lim_n I_\mu{}^*(f_n) < \infty$. Then $-\infty < I_\mu{}^*(f_1) \leqq I_\mu{}^*(f_n) < \infty$, hence f_n agrees a.e. with an integrable function g_n for each n. The monotone convergence theorem for I_μ applies to the sequence $\{g_n\}_n$ in $L_1(\mu)$ and consequently $I_\mu{}^*(\lim_n f_n) = I_\mu{}^*(\lim_n g_n) = \lim_n I_\mu(g_n) = \lim_n I_\mu{}^*(f_n)$. ■

3 Fatou's Lemma *If $\{f_n\}_n$ is a sequence in $L^*(\mu)$ which is bounded below a.e. by an integrable function g, then $I_\mu{}^*(\liminf_n f_n) \leqq \liminf_n I_\mu{}^*(f_n)$, and if $\{f_n\}_n$ is bounded above a.e. by an integrabl function, then $I_\mu{}^*(\limsup_n f_n) \geqq \limsup_n I_\mu{}^*(f_n)$.*

PROOF For $p > m$, $f_p \geqq \bigwedge_{k=m}^\infty f_k$ so that $I_\mu{}^*(f_p) \geqq I_\mu{}^*(\bigwedge_{k=m}^\infty f_k)$, and so $I_\mu{}^*(\bigwedge_{k=m}^\infty f_k) \leqq \liminf_p I_\mu{}^*(f_p)$ for each m. The sequence $\{\bigwedge_{k=m}^\infty f_k\}_m$ is increasing with pointwise limit, $\liminf_k f_k$, and $I_\mu{}^*(\bigwedge_{k=1}^\infty f_k) \geqq I_\mu(g) > -\infty$ so by the preceding theorem, $I_\mu{}^*(\liminf_k f_k) = \lim_m I_\mu{}^*(\bigwedge_{k=m}^\infty f_k) \leqq \liminf_p I_\mu{}^*(f_p)$.

Applying this conclusion to $\{-f_n\}_n$ yields the assertion about $\limsup_n f_n$. ■

A form of the dominated convergence theorem for $I_\mu{}^*$ follows from Fatou's lemma, just as for I_μ (see chapter 3). If $\{f_n\}_n$ is a sequence in $L^*(\mu)$ and g is an integrable function such that $|f_n| \leqq g$ a.e. for all n, then $I_\mu{}^*(\liminf_n f_n) \leqq \liminf I_\mu{}^*(f_n) \leqq \limsup_n I_\mu{}^*(f_n) \leqq I_\mu{}^*(\limsup_n f_n)$. Consequently, if $\liminf_n f_n = \limsup_n f_n$ a.e., the preceding inequalities must be equalities and $\{I_\mu{}^*(f_n)\}_n$ converges to $I_\mu{}^*(\liminf_n f_n) = I_\mu{}^*(\limsup_n f_n)$. (Both $\liminf_n f_n$ and $\limsup_n f_n$ belong to $L^*(\mu)$ in view of proposition 1.) This form of the dominated convergence theorem can also be derived directly from the earlier version in chapter 3.

Here is another example of the use of the B. Levi theorem for $I_\mu{}^*$. If μ is a regular measure on a δ-ring $\mathscr{A}$ of subsets of $\mathbb{R}$, then for A in $\mathscr{A}$ and for $e > 0$ there is in $\mathscr{A}$ a compact set K and an open set U such that $K \subset A \subset U$ and $\mu(U) < \mu(K) + e$. Then χ_K is upper semi-continuous, χ_U is lower semi-continuous, $\chi_K \leqq \chi_A \leqq \chi_U$ and $I_\mu{}^*(\chi_U) < I_\mu{}^*(\chi_K) + e$. By taking countable linear combinations, using the $e/2^n$ trick and Levi's theorem, one finds that any non-negative member of $L_1(\mu)$—and hence any member—can be approximated from above by lower semi-continuous members of $L^*(\mu)$ and from below by upper semi-continuous. Explicitly:

4 Proposition *If μ is a regular measure on a δ-ring of subsets of $\mathbb{R}$, $f \in L_1(\mu)$ and $e > 0$, then there are members s and t of $L^*(\mu)$, s upper semi-continuous and t lower semi-continuous, such that $s \leqq f \leqq t$ and $I_\mu{}^*(s) + e > I_\mu{}^*(t)$.*

Proposition 4 furnishes a link between our treatment of integration and the classical Daniell treatment of the Lebesgue integral. A lower semi-continuous function f on $[0:1]$ to $(-\infty:\infty]$ is the pointwise limit of an increasing sequence $\{f_n\}_n$ of continuous real valued functions on $[0:1]$, and the Lebesgue integral* of f is just the supremum of the Riemann integrals of the f_n. In a similar fashion, one can describe the Lebesgue integral of an upper semi-continuous function f on $[0:1]$ to $[-\infty:\infty)$ as the infimum of the Riemann integrals of continuous functions above f. The preceding proposition leads to a description of an arbitrary Lebesgue integrable function in terms of approximation *from above* by l.s.c. $(-\infty:\infty]$ valued functions and *from below* by u.s.c. $[-\infty:\infty)$ valued functions. It is worth noticing that one cannot always approximate a Lebesgue integrable function f on $[0:1]$ from above by an u.s.c. function (e.g., the characteristic function of the set of rational numbers in $[0:1]$), nor by a real, *finite valued*, l.s.c. function (e.g., $f(x) = 0$ for x irrational, $f(p/q) = q$ for p and q relatively prime positive integers with $q \geqq p$).

We now use the results on integrable* functions to construct a measure for $X \times Y$ from measures for X and for Y. We suppose throughout that **μ is a measure on a δ-ring $\mathscr{A}$ of subsets of X and ν is a measure on a σ-ring $\mathscr{B}$ of subsets of Y.** For each function f on $X \times Y$ to $\mathbb{R}^*$ and for each (x, y) in $X \times Y$, the **horizontal section of f** through y, f^y, is the function $x \mapsto f(x, y)$ and the **vertical section** through x, f_x, is $y \mapsto f(x, y)$. Thus $f_x(y) = f(x, y) = f^y(x)$ for all (x, y) in $X \times Y$.

If $f^y \in L^*(\mu)$ for each y, then $E_\mu(f)$ is the function $y \mapsto I_\mu^*(f^y) = \int f(x, y)\, d\mu x$, and if $f_x \in L^*(\nu)$ for each x, then $E^\nu(f)$ is the function $x \mapsto I_\nu^*(f_x) = \int f(x, y)\, d\nu y$. If $E_\mu(f) \in L^*(\nu)$, then $I_\nu^* \circ E_\mu(f)$ is an **iterated integral** of f, denoted $\int\int f(x, y)\, d\mu x\, d\nu y$, and if $E^\nu(f) \in L^*(\mu)$, then $\int\int f(x, y)\, d\nu y\, d\mu x = I_\mu^* \circ E^\nu(f)$. We deduce from theorem 2 that if $\{f_n\}_n$ is a sequence of non-negative functions on $X \times Y$ to $\mathbb{R}^*$ such that $f_n^y \in L^*(\mu)$ for all n and y and $E_\mu(f_n) \in L^*(\nu)$ for each n, then $(\sum_n f_n)^y \in L^*(\mu)$ for each y, $E_\mu(\sum_n f_n) \in L^*(\nu)$ and $I_\nu^* \circ E_\mu(\sum_n f_n) = \sum_n I_\nu^* \circ E_\mu(f_n)$.

A function f on $X \times Y$ to $\mathbb{R}$ is **compatible with μ and ν** iff $x \mapsto f(x, y)$ is μ integrable for each y, $y \mapsto \int f(x, y)\, d\mu x$ is ν integrable, $y \mapsto f(x, y)$ is ν integrable for each x, $x \mapsto \int f(x, y)\, d\nu y$ is μ integrable, and $\int\int f(x, y)\, d\mu x\, d\nu y = \int\int f(x, y)\, d\nu y\, d\mu x$. A subset D of $X \times Y$ is **compatible with μ and ν** iff its characteristic function is compatible.

5 Lemma on Compatibility *The family of compatible sets is closed under disjoint union, proper difference, and disjoint dominated countable union.*

If μ and ν are complete measures, A is a compatible set, $\int\int \chi_A(x, y)\, d\mu x\, d\nu y = 0$ and $B \subset A$ then B is compatible and $\int\int \chi_B(x, y)\, d\mu x\, d\nu y = 0$.

PROOF The family F of compatible functions is evidently a vector space. Suppose that A and B are compatible sets. If A and B are disjoint, then $\chi_{A\cup B} = \chi_A + \chi_B \in F$ so $A \cup B$ is compatible, and if $B \subset A$, then $\chi_{A\setminus B} = \chi_A - \chi_B \in F$ so $A\setminus B$ is compatible. If $\{B_n\}_n$ is a disjoint sequence of compatible subsets of a compatible set A and $B = \bigcup_n B_n$, then $\chi_B = \sum_n \chi_{B_n} \leqq \chi_A$ so $I_\mu^* \circ E^\nu(\chi_B) = \sum_n I_\mu^* \circ E^\nu(\chi_{B_n}) \leqq I_\mu^* \circ E^\nu(\chi_A) < \infty$. It follows without difficulty that B is compatible, and the first assertion of the lemma is established. The second assertion is a consequence of the fact that, if μ and ν are complete measures, then both iterated integrals of χ_A vanish iff for μ a.e. x, $\chi_A(x, y)$ vanishes for ν a.e. y, and for ν a.e. y, $\chi_A(x, y)$ vanishes for μ a.e. x, because if χ_A satisfies this condition then so does χ_B for every subset B of A. ■

If $A \in \mathscr{A}$ and $B \in \mathscr{B}$, then the rectangle $A \times B$ is a compatible set, since $\chi_{A\times B}(x, y) = \chi_A(x)\chi_B(y)$, so $\int\int \chi_{A\times B}(x, y)\, d\mu x\, d\nu y = \mu(A)\nu(B) = \int\int \chi_{A\times B}(x, y)\, d\nu y\, d\mu x$. The **product δ-ring $\mathscr{A} \otimes \mathscr{B}$** is defined to be the δ-ring generated by the family of all rectangles $A \times B$ with A in $\mathscr{A}$ and B in $\mathscr{B}$. The intersection of two rectangles is a rectangle, and consequently theorem 4.6 shows that $\mathscr{A} \otimes \mathscr{B}$ is the smallest family containing all rectangles that is closed under disjoint union, proper difference, and disjoint dominated countable union. The preceding lemma then implies that every member of $\mathscr{A} \otimes \mathscr{B}$ is a compatible set, so the two iterated integrals of χ_C agree and are finite for each C in $\mathscr{A} \otimes \mathscr{B}$.

For C in $\mathscr{A} \otimes \mathscr{B}$, let **$\mu \otimes \nu(C)$** be the iterated integral $\int\int \chi_C(x, y)\, d\mu x\, d\nu y$. The iterated integral is countably additive whence $\mu \otimes \nu$ is a non-negative, real valued, countably additive function on $\mathscr{A} \otimes \mathscr{B}$. Thus $\mu \otimes \nu$ is a measure, the **product measure** of μ and ν. It is the unique measure on the δ-ring generated by rectangles $A \times B$, A in $\mathscr{A}$ and B in $\mathscr{B}$, such that $\mu \otimes \nu(A \times B) = \mu(A)\nu(B)$.

Let us agree that a set A is **null compatible** with μ and ν iff A is compatible and the iterated integrals of χ_A are zero. Evidently each set of $\mu \otimes \nu$ measure zero is null compatible. If μ and ν are complete measures, then each subset of a null compatible set is of the same sort by lemma 5. It follows that members of $(\mathscr{A} \otimes \mathscr{B})^\vee$, where $(\mu \otimes \nu)^\vee$ on $(\mathscr{A} \otimes \mathscr{B})^\vee$ denotes the usual completion of $\mu \otimes \nu$, consists of compatible sets. We record these facts after adopting some notation.

For each subset C of $X \times Y$ and each member (x, y), the **horizontal section of C through y, ${}^yC$,** is $\{u: (u, y) \in C\}$ and the **vertical section of C through x, ${}_xC$,** is $\{v: (x, v) \in C\}$. If C is compatible with μ and ν, then $\int \chi_C(x, y)\, d\mu x = \mu({}^yC)$ and $\int \chi_C(x, y)\, d\nu y = \nu({}_xC)$.

6 THEOREM ON SECTIONS *If $C \in \mathscr{A} \otimes \mathscr{B}$, then the product integral $I_{\mu\otimes\nu}$ agrees with both iterated integrals at χ_C, so $\mu \otimes \nu(C) = \int \mu({}^yC)\, d\nu y = \int \nu({}_xC)\, d\mu x$.*

If μ and ν are complete measures, then the integral w.r.t. the usual completion of $\mu \otimes \nu$ agrees with the iterated integral at χ_C for C in the completion $(\mathscr{A} \otimes \mathscr{B})^\vee$.

7 REMARKS

(i) Product measure $\mu \otimes \nu$ is the unique measure on $\mathscr{A} \otimes \mathscr{B}$ such that $\mu \otimes \nu(A \times B) = \mu(A)\nu(B)$ for all rectangles $A \times B$ with A in $\mathscr{A}$ and B in $\mathscr{B}$. The intersection of two rectangles is a rectangle and the difference of two rectangles is the union of finitely many disjoint rectangles. Consequently the family $\mathscr{R}$ of such unions is a ring of sets and so $\mu \otimes \nu | \mathscr{R}$ is an exact pre-measure, and this pre-measure induces $\mu \otimes \nu$. Consequently each member of $\mathscr{A} \otimes \mathscr{B}$ has inner approximations in $\mathscr{R}_\delta$ and outer approximations in $\mathscr{R}_\sigma$.

(ii) The usual completion of a measure ρ on a δ-ring $\mathscr{D}$ is obtained by letting $\mathscr{N}$ be the family of all subsets of sets of ρ measure zero, and assigning measure $\rho(D)$ to each symmetric difference $D \,\Delta\, N$ with D in $\mathscr{D}$ and N in $\mathscr{N}$. The only requirements on the family $\mathscr{N}$ needed to ensure that this process yields a "completion" of ρ is that each subset of a member of $\mathscr{N}$ belong to $\mathscr{N}$, $\mathscr{N}$ is closed under Δ, and $\mathscr{N} \cap \mathscr{D} = \{A\colon A \in \mathscr{D} \text{ and } \rho(A) = 0\}$.

One can construct different completions by using different families $\mathscr{N}$. For example, the family of null sets and the family of locally null sets yield completions which are convenient for certain purposes. Every completion of $\mathscr{N}$ is evidently an extension of the usual completion. If μ and ν are complete measures, $\rho = \mu \otimes \nu$ and $\mathscr{N}$ is the family of all subsets of sets that are null compatible with μ and ν, then a completion of $\mu \otimes \nu$ is obtained, the **null compatible completion**, that agrees with the iterated integrals on each member of its domain. It is the "largest" completion with this property.

According to theorem 6, both of the iterated integrals are extensions of $I_{\mu \otimes \nu} | \{\chi_C\colon C \in \mathscr{A} \otimes \mathscr{B}\}$. Since $I_{\mu \otimes \nu}$ and both interated integrals are countably additive, we infer that $I_{\mu \otimes \nu}$ agrees with the iterated integrals on the class of all linear combinations $\sum_n c_n \chi_{C_n}$ with $\{c_n\}_n$ in $\mathbb{R}^+$ and $\{C_n\}_n$ in $\mathscr{A} \otimes \mathscr{B}$. This class $\boldsymbol{L^+(\mathscr{A} \otimes \mathscr{B})} = \{f\colon f \textit{ is non-negative, } \mathbb{R}^* \textit{ valued and } \mathscr{A} \otimes \mathscr{B} \ \sigma\textit{-simple}\}$ is identical with the class of non-negative $I_{\mu \otimes \nu}{}^*$ integrable functions. This establishes the following.

8 TONELLI THEOREM *If μ and ν are measures, $\mu \otimes \nu$ is their product measure, and $f \in L^+(\mathscr{A} \otimes \mathscr{B})$, then $\int f \, d\mu \otimes \nu = \int (\int f(x, y) \, d\nu y) \, d\mu x$. In detail: $f_x \in L^*(\nu)$ for each x in X, $E^\nu(f) \in L^*(\mu)$ and $I_{\mu \otimes \nu}{}^*(f) = I_\mu{}^* \circ E^\nu(f) = I_\nu{}^* \circ E_\mu(f)$.*

Moreover; if μ and ν are complete and $f \in L^+((\mathscr{A} \otimes \mathscr{B})^\vee)$, then $I_{(\mu \otimes \nu)^\vee}{}^*(f) = I_\mu{}^* \circ E^\nu(f) = I_\nu{}^* \circ E_\mu(f)$.

The Fubini theorem is a corollary to the Tonelli theorem. It is convenient, before stating the theorem, to make the **convention** that if f is a function that agrees μ a.e. with a member g of $L_1(\mu)$ then $\int f(x)\,d\mu x$ or $\int f\,d\mu$, is $\int g\,d\mu$. (*We are surreptitiously enlarging the family of functions that we can "integrate".*)

9 FUBINI THEOREM *Let μ and ν be measures, let $\mu \otimes \nu$ be the product measure and let $(\mu \otimes \nu)^\vee$ be its completion. Then $\int f(x, y)\,d\mu \otimes \nu(x, y) = \int(\int f(x, y)\,d\nu y)\,d\mu x$ for all f in $L_1(\mu \otimes \nu)$, and also for f in $L_1((\mu \otimes \nu)^\vee)$ provided ν is complete.*

In detail: if $f \in L_1(\mu \otimes \nu)$, or if $f \in L_1(\mu \otimes \nu)^\vee)$ and ν is complete, then $f_x \in L_1(\nu)$ for μ a.e. x, and if $h(x) = I_\nu(f_x)$ for such x, then h agrees μ a.e. with a member g of $L_1(\mu)$ and $\int g\,d\mu = \int f\,d\mu \otimes \nu$.

PROOF Suppose $f \in L_1(\mu \otimes \nu)$ and $f \geqq 0$. Then $I_{\mu \otimes \nu}(f) = I_{\mu \otimes \nu}{}^*(f) = I_\mu{}^* \circ E^\nu(f) < \infty$, and consequently $E^\nu(f)(x)$ is finite except for x in some μ null set D. If $x \notin D$, then $E^\nu(f)(x) = I_\nu{}^*(f_x)$ is finite, so $f_x \in L_1(\nu)$. Thus $\int f(x, y)\,d\nu y$ is defined and agrees with $E^\nu(f)(x)$ for μ a.e. x, so $\int(\int f(x, y)\,d\nu y)\,d\mu x = I_\mu{}^* \circ E^\nu(f) = I_{\mu \otimes \nu}(f)$.

The same line of argument establishes the theorem for a member of $L_1((\mu \otimes \nu)^\vee)$, provided ν is complete. ■

10 COROLLARY *Both iterated integrals agree with $I_{\mu \otimes \nu}$ on $L_1(\mu \otimes \nu)$, and if μ and ν are complete then the iterated integrals agree with $I_{(\mu \otimes \nu)^\vee}$ on $L_1((\mu \otimes \nu)^\vee)$.*

A particular consequence of the preceding is that one "may interchange the order of integration if the integrand is integrable with respect to the product measure". One might suppose that the two iterated integrals are *always* equal, but this would be a hazardous supposition. Here are two examples.

Let both μ and ν be counting measures for the set $\mathbb{N}$ of natural numbers and let $f(m, n)$ be 1 if $m = n$ and -1 if $n = m + 1$ and zero elsewhere. Then one of the iterated integrals is one and the other is zero. This is essentially an "infinity minus infinity" trouble.

Here is an example of a different character. Let μ be any measure (like $\Lambda_{[a:b]}$) on the Borel subsets of $[0:1]$ such that $\mu([0:1]) = 1$ and $\mu(\{x\}) = 0$ for each x in $[0:1]$, let ν be counting measure for $[0:1]$, and let f be the characteristic function of the diagonal $\{(x:x): x \in [0:1]\}$. Then, again, one iterated integral is one and the other is zero. This is a "borel measurability problem".

Finally, the "almost everywhere" qualifications in the statement of the Fubini theorem are essential. It is easy to define $\mu \otimes \nu$ integrable functions f such that f_x is not always ν integrable (e.g., for Lebesgue–Borel measure $\Lambda \otimes \Lambda$ for the plane, the function that sends (x, y) to

$1/\sqrt{x^2 + y^2}$ for $0 < x^2 + y^2 < 1$ and into 0 otherwise). It is also easy to define bounded non-negative functions f such that f_x is always ν integrable but $x \mapsto \int f(x, y)\, d\nu x$ is not always μ integrable.

SUPPLEMENT: BOREL PRODUCT MEASURE

The product of two regular Borel measures may fail to be a Borel measure, but it has a natural extension which is a regular Borel measure, and the Tonelli and Fubini theorems extend to this measure. We establish these facts after two preliminary lemmas.

If λ is a regular Borel measure for a locally compact Hausdorff space Z and f is a bounded member of $L_1(\lambda)$, then f can be approximated from above by a lower semi-continuous member of $L_1(\lambda)$, and from below by a u.s.c. member, because proposition 4 and its proof apply directly to regular Borel measures. Moreover, a bounded, non-negative u.s.c. function g that vanishes outside a compact set K is necessarily λ integrable because: it is locally $\mathscr{B}^\delta(Z)$ measurable and has compact support and hence is $\mathscr{B}^\delta(Z)$ σ-simple, and since $g \leqq b\chi_K \in L_1(Z)$, $I_\lambda^*(g) < \infty$. We use these facts to establish a "hypercontinuity" property for I_λ on the class of such functions g.

11 LEMMA *Suppose λ is a regular Borel measure for Z, K is a compact subset of Z, $\{f_\alpha\}_{\alpha \in D}$ is a decreasing net of real valued, non-negative upper semi-continuous functions vanishing outside K and $f = \inf_\alpha f_\alpha$.*

Then each f_α and f are λ integrable and $I_\lambda(f) = \inf_{\alpha \in D} I_\lambda(f_\alpha)$.

PROOF The function f is u.s.c. because it is the limit of a decreasing net of u.s.c. functions, and each real valued u.s.c. function on a compact set is bounded because it assumes its supremum. It follows that each f_α and f are λ integrable. Evidently $I_\lambda(f) \leqq \mathit{inf}_{\alpha \in D} I_\lambda(f_\alpha)$, and the reverse inequality remains to be proved.

For $e > 0$ choose a real valued l.s.c. function g such that $g \geqq f$ and $I_\lambda(g - f) < e$. Then $\{(f_\alpha - g) \vee 0\}_{\alpha \in D}$ is a decreasing net of u.s.c. functions converging pointwise, and so uniformly, to zero (observe, with Dini: if $d > 0$, then $\bigcap_{\alpha \in D}\{z: f_\alpha(z) - g(z) > d\} = \varnothing$). Consequently, for α large, $f_\alpha(z) - g(z) < e$ for all z, whence $\mathit{inf}_{\alpha \in D} I_\lambda(f_\alpha) < I_\lambda(g) + e\lambda(K) < I_\lambda(f) + e + e\lambda(K)$. The desired equality follows. ∎

We assume for the remainder of this supplement that **μ and ν are regular Borel measures for locally compact Hausdorff spaces X and Y respectively**.

The following lemma will help establish that each compact subset A of $X \times Y$ is **compatible** with μ and ν; that is, for each (x, y) in $X \times Y$, the vertical section $(\chi_A)_x \in L_1(\nu)$ and the horizontal section $(\chi_A)^y \in L_1(\mu)$, the function E^ν given by $E^\nu(\chi_A)(x) = \int \chi_A(x, y)\, d\nu y = \nu({}_xA)$ is μ integrable, E_μ is ν integrable, and the iterated integrals $I_\mu \circ E^\nu$ and $I_\nu \circ E_\mu$ are equal.

12 LEMMA *If $\{A_\alpha\}_{\alpha \in D}$ is a decreasing net of compact sets, each of which is compatible with μ and ν and $A = \bigcap_{\alpha \in D} A_\alpha$, then A is compatible and $\iint \chi_A(x, y)\, d\nu y\, d\mu x = \lim_{\alpha \in D} \iint \chi_{A_\alpha}(x, y)\, d\nu y\, d\mu x$.*

PROOF For each x in X, the vertical sections $\{(\chi_{A_\alpha})_x\}_{\alpha \in D}$ form a decreasing net of u.s.c. functions on Y so the preceding lemma implies that $\nu({}_xA) = \int \chi_A(x, y)\, d\nu y = lim_{\alpha \in D} \int \chi_{A_\alpha}(x, y)\, d\nu y = lim_{\alpha \in D}\, \nu({}_xA_\alpha)$. That is, the pointwise limit of $\{E^\nu(\chi_{A_\alpha})\}_{\alpha \in D}$ is $E^\nu(\chi_A)$.

For each compact subset C of $X \times Y$, the function $x \mapsto \nu({}_xC)$ is u.s.c. for the following reasons. For x in X and $e > 0$ there is an open member V of $\mathscr{B}^\delta(Y)$ so ${}_xC \subset V$ and $\nu(V) < \nu({}_xC) + e$ because ν is regular. Because C is compact there is a neighborhood U of x such that ${}_uC \subset V$ for u in U, whence $\nu({}_uC) \leqq \nu(V) < \nu({}_xC) + e$. Thus the function $x \mapsto \nu({}_xC)$, which is $E^\nu(\chi_C)$, is upper semi-continuous.

Since the decreasing net $\{E^\nu(\chi_{A_\alpha})\}_{\alpha \in D}$ of u.s.c. functions converges pointwise to $E^\nu(\chi_A)$, $I_\mu \circ E^\nu(\chi_A) = lim_{\alpha \in D} I_\mu \circ E^\nu(\chi_{A_\alpha})$ by lemma 11, and $I_\nu \circ E_\mu(\chi_A) = lim_{\alpha \in D} I_\nu \circ E_\mu(\chi_{A_\alpha})$ in similar fashion. But for each α, the two iterated integrals of χ_{A_α} are equal, and so this is also the case for the limit function χ_A. ■

If D and E are compact subsets of X and Y respectively, then the rectangle $D \times E$ is compatible with μ and ν and the intersection of two such rectangles is such a rectangle. Consequently every union of finitely many compact rectangles is compatible with μ and ν. Each neighborhood of a compact subset A of $X \times Y$ contains a neighborhood that is a finite union of compact rectangles and the family of such neighborhoods is directed by $\subset$. Lemma 12 then implies that A is compatible with μ and ν.

The family of compatible subsets of $X \times Y$ is closed under disjoint union, proper difference and dominated disjoint countable union according to lemma 5, and since the intersection of two compact sets is compact, theorem 4.6 on generated δ-rings implies that every member of $\mathscr{B}^\delta(X \times Y)$ is compatible with μ and ν. The iterated integral is countably additive and so $\boldsymbol{\mu \otimes_{\mathscr{B}} \nu}$, where $\mu \otimes_{\mathscr{B}} \nu(B)$ is the iterated integral of χ_B for each B in $\mathscr{B}^\delta(X \times Y)$, is a measure. It is called the **Borel product measure**. This measure $\mu \otimes_{\mathscr{B}} \nu$ is an extension of the product measure $\mu \otimes \nu$ since the value of each at B is the iterated integral of the characteristic function of B.

For convenience let $\mathscr{N}$ be $\{A: A \in \mathscr{B}^\delta(X \times Y)$ *and* $\mu \otimes_{\mathscr{B}} \nu(A) = 0\}$. Here are some important properties of the Borel product.

13 THEOREM ON BOREL PRODUCT MEASURE *Each member B of $\mathscr{B}^\delta(X \times Y)$ is compatible with μ and ν, and each iterated integral of χ_B is $\mu \otimes_{\mathscr{B}} \nu(B)$. The Borel product $\mu \otimes_{\mathscr{B}} \nu$ is a regular Borel measure that extends $\mu \otimes \nu$ and it is the unique regular Borel measure that extends $D \times E \mapsto \mu(D)\nu(E)$ for compact rectangles $D \times E$.*

The δ-ring $\mathscr{B}^\delta(X \times Y)$ is the family of symmetric differences $B \triangle N$ with B in $\mathscr{B}^\delta(X) \otimes \mathscr{B}^\delta(Y)$ and N in $\mathscr{N}$.

PROOF It follows from lemma 12 and the definition of $\mu \otimes_{\mathscr{B}} \nu$ that for each compact subset A of $X \times Y$, $\mu \otimes_{\mathscr{B}} \nu(A)$ is the infimum of $\mu \otimes_{\mathscr{B}} \nu(B)$ for B a compact neighborhood of A, and consequently $\mu \otimes_{\mathscr{B}} \nu$ restricted to the family of compact subsets is a regular content and hence (theorem 4.15) extends to a regular Borel measure. This measure agrees with $\mu \otimes_{\mathscr{B}} \nu$ on compacta and therefore also on the generated δ-ring $\mathscr{B}^\delta(X \times Y)$, so $\mu \otimes_{\mathscr{B}} \nu$ is regular.

A regular Borel measure that agrees with $\mu \otimes_{\mathscr{B}} \nu$ on compact rectangles also agrees on the family of finite unions of compact rectangles, hence by regularity on all compacta, and consequently is identical with $\mu \otimes_{\mathscr{B}} \nu$.

The family of all symmetric differences $B \triangle N$ with B in $\mathscr{B}^\delta(X) \otimes \mathscr{B}^\delta(Y)$ and N in $\mathscr{N}$ is a δ-ring and a subfamily of $\mathscr{B}^\delta(X \times Y)$. We show that each compact set A, and hence each member of $\mathscr{B}^\delta(X \times Y)$ belongs to the subfamily and this will establish the last assertion of the theorem.

Each compact set A is the intersection of compact neighborhoods B of A such that $B \in \mathscr{B}^\delta(X) \otimes \mathscr{B}^\delta(Y)$, nd $\mu \otimes_{\mathscr{B}} \nu(A)$ is the infimum of $\mu \otimes_{\mathscr{B}} \nu(B)$ for these neighborhoods B. There is then a sequence $\{B_n\}_n$ of such neighborhoods such that $\mu \otimes_{\mathscr{B}} \nu(A) = lim_n \mu \otimes_{\mathscr{B}} \nu(B_n)$. Let $B = \bigcap_n B_n$ and $N = B \backslash A$. Then $B \in \mathscr{B}^\delta(X) \otimes \mathscr{B}^\delta(Y)$, $N \in \mathscr{B}^\delta(X \times Y)$, $\mu \otimes_{\mathscr{B}} \nu(N) = 0$ and $A = B \triangle N$. ∎

The last statement of the preceding theorem was established by first proving that each compact subset A of $X \times Y$ is a subset of a compact member B of $\mathscr{B}^\delta(X) \otimes \mathscr{B}^\delta(Y)$ such that $\mu \otimes_{\mathscr{B}} \nu(B \backslash A) = 0$. It is worth noticing that each open member U of $\mathscr{B}^\delta(X \times Y)$ contains an open member V of $\mathscr{B}^\delta(X) \otimes \mathscr{B}^\delta(Y)$ such that $\mu \otimes_{\mathscr{B}} \nu(U \backslash V) = 0$ (let $A = U^- \backslash U$, and use the foregoing result).

Both the iterated integrals and the extended integral w.r.t. $\mu \otimes_{\mathscr{B}} \nu$ are countably additive on the class L^+ of non-negative $\mathbb{R}^*$ valued $\mathscr{B}^\delta(X \times Y)$ σ-simple functions, and since they agree on characteristic functions of members of $\mathscr{B}^\delta(X \times Y)$, they agree on L^+. The class L^+ can also be described as the family of non-negative $\mathscr{L}\mathscr{B}^\delta(X \times Y)$ measurable functions that vanish off a countable union of members of $\mathscr{B}^\delta(X \times Y)$, and since each such countable union is a subset of a countable union of compacta, L^+ is just the family of non-negative locally

Borel measurable functions with σ-compact support. This establishes the Tonelli theorem for $\mu \otimes_{\mathscr{B}} \nu$ and the Fubini theorem follows from it.

The usual completion of a measure is obtained by enlarging the family of sets of measure zero, by adjoining all subsets of such sets. The usual completion of a regular Borel measure is both inner and outer regular but its domain generally includes some non-Borel sets. We leave to the reader the proof that the Tonelli and Fubini theorems hold for the usual completions of μ, ν and $\mu \otimes_{\mathscr{B}} \nu$.

Here is a summary of the principal foregoing results. (Caution: the statement of the Fubini theorem presumes the convention on $\int\int$ that is made before theorem 9.)

14 SUMMARY *Let μ and ν be regular Borel measures for locally compact Hausdorff spaces and let $\mu \otimes_{\mathscr{B}} \nu$ be their Borel product.*

(i) (TONELLI) *If f is a $\mathbb{R}^*$ valued non-negative locally Borel measurable function with σ-compact support, then so are the functions $(x \mapsto \int f(x, y)\, d\nu y) \in L^*(\mu)$, $(y \mapsto \int f(x, y)\, d\mu x) \in L^*(\nu)$ for each (x, y) in $X \times Y$, and the two iterated integrals agree with the extended integral w.r.t. $\mu \otimes_{\mathscr{B}} \nu$.*

(ii) (FUBINI) *If $f \in L_1(\mu \otimes_{\mathscr{B}} \nu)$, then $\int f\, d\mu \otimes_{\mathscr{B}} \nu = \int\int f(x, y)\, d\mu x\, d\nu y = \int\int f(x, y)\, d\nu y\, d\mu x$.*

(iii) *Both the Tonelli and Fubini theorems hold for the usual completions of μ, ν and $\mu \otimes_{\mathscr{B}} \nu$.*

15 NOTES The Fubini theorem for Borel product of regular measures has been widely assumed but the first proofs available in the literature are due to W. W. Bledsoe and A. P. Morse, *Trans. Amer. Math. Soc.* **79** (1955), 173–215, MR 16, 1008; Karel de Leeuw, *Math. Scand.* **11** (1962), 117–122, MR 33, 4179; and M. Hable and M. Rosenblatt, *Proc. Amer. Math. Soc.* **14** (1963), 177–184, MR 30, 214. A close examination of the Borel product and stronger versions of the Fubini theorem are provided by Roy A. Johnson, *Trans. Amer. Math. Soc.* **123** (1966), 112–129, MR 33, 5832.

Chapter 8

MEASURES* AND MAPPINGS

A **measure in the extended sense**, or just a **measure***, is a non-negative, countably additive, $\mathbb{R}^*$ valued function μ on a δ-ring $\mathscr{A}$ with $\mu(\varnothing) = 0$. The function on $\mathscr{A}$ that is 0 at $\varnothing$ and ∞ elsewhere is a measure*, each measure is a measure*, and each finite valued measure* is a measure. Classical Lebesgue measure for $\mathbb{R}$ (see note 4.13(i)) is the prototypical example of a measure*. A function f is **integrable** (or **integrable***) **w.r.t. a measure*** μ on $\mathscr{A}$ iff it is integrable (integrable*) w.r.t. the measure $\mu_0 = \mu|\{A: A \in \mathscr{A} \text{ and } \mu(A) < \infty\}$ and in this case $\int f\,d\mu = \int f\,d\mu_0$. Thus the integral w.r.t. classical Lebesgue measure is indentical with the integral w.r.t. Λ^1.

Each measure μ on a δ-ring $\mathscr{A}$ can be extended to a measure* on the σ-field $\mathscr{L}\mathscr{A}$ of locally $\mathscr{A}$ measureable sets, usually in many different ways. Here is an extreme example: if $X = \mathbb{R}$, $\mathscr{A} = \{\varnothing, \{1\}\}$ and $\mu(\{1\}) = 0$, then $\mathscr{L}\mathscr{A}$ is the class $\mathscr{P}(\mathbb{R})$ of all subsets of $\mathbb{R}$, and every measure* ν on $\mathscr{P}(\mathbb{R})$ for which $\nu(\{1\}) = 0$ is an extension of μ.

The **canonical extension $\mu_\#$ of a measure μ on $\mathscr{A}$—or of a measure* μ**—is defined by setting $\mu_\#(B) = sup\{\mu(A): A \in \mathscr{A} \text{ and } A \subset B\}$ for each B in $\mathscr{L}\mathscr{A}$. Evidently $\mu_\#$ is an extension of μ and each measure* on $\mathscr{L}\mathscr{A}$ that extends μ is greater than or equal to $\mu_\#$. It is straightforward to check that $\mu_\#$ is countably additive, and so $\mu_\#$ is the minimal measure* on $\mathscr{L}\mathscr{A}$ that extends μ. Classical Lebesgue measure is the canonical extension of Λ^1.

Not every measure* on $\mathscr{L}\mathscr{A}$ is the canonical extension of a measure, because each such canonical extension $\mu_\#$ has the property: $\mu_\#(B) = sup\{\mu_\#(A): A \in \mathscr{L}\mathscr{A}, A \subset B \text{ and } \mu_\#(A) < \infty\}$ for B in $\mathscr{L}\mathscr{A}$. Such measures* are sometimes called *semi-finite*. The measure* $\mu_\#$ is finite valued (and thus a measure) iff μ is a bounded function on $\mathscr{A}$; that is, μ is **totally finite**. In this case $\mu_{\#}(X)$ is the **total μ mass.**

It is not difficult to describe the integral w.r.t. $\mu_{\#}$ in terms of the integral w.r.t. μ. If $B \in \mathscr{L}\mathscr{A}$ and $\mu_{\#}(B) < \infty$, then there is an increasing sequence $\{A_n\}_n$ in $\mathscr{A}$ such that $A = \bigcup_n A_n \subset B$ and $\mu_{\#}(B) = lim_n\, \mu(A_n)$. If $B \in \mathscr{A}_\sigma$, we may assume $A = B$, but in any case $\mu_{\#}(B\backslash A) = 0$, whence $B\backslash A$ is **locally of μ measure zero** in the sense that $\mu(C \cap (B\backslash A)) = 0$ for all C in $\mathscr{A}$. Thus $\chi_{B\backslash A} = 0$ **locally μ a.e.** (that is, except on a subset of a set of locally μ measure zero). Hence χ_B is $\mu_{\#}$ integrable iff $B \in \mathscr{L}\mathscr{A}$ and for some subset D of B, χ_D is μ integrable and $\chi_D = \chi_B$ locally μ almost everywhere. If $B \in \mathscr{A}_\sigma$, then $\chi_{B\backslash D}$ is itself μ integrable, and consequently so is χ_B. This establishes the following proposition for characteristic function of members of $\mathscr{L}\mathscr{A}$, and it extends directly to linear combinations with non-negative coefficients of countably many such functions.

1 PROPOSITION *A real valued function f is integrable w.r.t. the canonical extension $\mu_{\#}$ of a measure μ on $\mathscr{A}$ iff it is $\mathscr{L}\mathscr{A}$ measurable and agrees locally μ a.e. with some μ integrable function g, and in this case $\int f\, d\mu_{\#} = \int g\, d\mu$.*

A $\mu_{\#}$ integrable function is μ integrable iff it has a support in $\mathscr{A}_\sigma$.

NOTE There is no difficulty in establishing this and several other propositions in this section for functions that are integrable*—that is, integrable in the extended sense. But we give here only the simplest forms of the results, since the propositions for integrable* functions are straightforward consequences of these.

Suppose μ is a measure* on a δ-ring $\mathscr{A}$ of subsets of X, $\mathscr{B}$ is a δ-ring of subsets of Y, and $T: X \to Y$ is a map such that $T^{-1}[B] \in \mathscr{L}\mathscr{A}$ for each B in $\mathscr{B}$. Then the **T image measure* of μ on $\mathscr{B}$**, denoted $T_{\mathscr{B}}\mu$ or just $T\mu$ if confusion is unlikely, is defined by setting $T\mu(B) = sup\{\mu(A): A \in \mathscr{A}$ and $A \subset T^{-1}[B]\}$ for each B in $\mathscr{B}$. If μ is a measure, then $T\mu(B)$ is just $\mu_{\#}(T^{-1}[B])$.

2 MAPPING LEMMA *Suppose $\mathscr{A}$ and $\mathscr{B}$ are δ-rings of subsets of X and Y respectively, μ is a measure on $\mathscr{A}$, $T: X \to Y$ is a map such that $T^{-1}[B] \in \mathscr{L}\mathscr{A}$ for all B in $\mathscr{B}$, and f is a real valued $\mathscr{B}$ σ-simple function on Y.*

Then $f \in L_1(T\mu)$ iff $f \circ T \in L_1(\mu_{\#})$, or iff $f \circ T$ agrees locally μ a.e. with a member of $L_1(\mu)$, and in this case $\int f\, dT\mu$ is $\int f \circ T\, d\mu_{\#}$, which is equal to $\int f \circ T\, d\mu$ iff $f \circ T$ has a support in $\mathscr{A}_\sigma$.

PROOF If $B \in \mathscr{B}$ and $T\mu(B) < \infty$ then $T^{-1}[B] \in \mathscr{L}\mathscr{A}$ and $\mu_{\#}(T^{-1}[B]) < \infty$, so $\int \chi_B\, dT\mu = \int \chi_{T^{-1}[B]}\, d\mu_{\#} = \int \chi_B \circ T\, d\mu_{\#}$. If f is a non-negative $\mathscr{B}$ σ-simple function on Y, then $f = \sum_n b_n \chi_{B_n}$ for some sequences $\{B_n\}_n$ in $\mathscr{B}$ and $\{b_n\}_n$ in $\mathbb{R}^+$, $f \circ T = \sum_n b_n \chi_{B_n} \circ T$, and $\int f \, dT\mu = \sum_n b_n T\mu(B_n) = \sum_n b_n \mu_\#(T^{-1}[B_n]) = \int f \circ T \, d\mu_\#$. This fact, together with proposition 1, establishes the lemma. ■

Here is a simple example of the use of the mapping lemma. Suppose $a \in \mathbb{R}$ and T is translation by a, in the sense that $T(x) = x + a$ for all x in $\mathbb{R}$. Borel–Lebesgue measure Λ is on $\mathscr{B} = \mathscr{B}^\delta(\mathbb{R})$ and is invariant under translation, so $T_\mathscr{B}\Lambda = \Lambda$. Consequently, by the mapping lemma $\int f \, d\Lambda = \int f \, dT_\mathscr{B}\Lambda = \int f \circ T \, d\Lambda = \int f(x + a) \, d\Lambda x$ for each Λ integrable function f. The same sort of equality holds for Lebesgue measure.

Here is another consequence of the mapping lemma. We are concerned with the image of a measure μ on $\mathscr{A}$ under a real valued map g. If $g^{-1}[B]$ is locally $\mathscr{A}$ measurable for each B in $\mathscr{B} = \mathscr{B}^\delta(\mathbb{R})$, and in particular if g is $\mathscr{A}$ σ-simple, then $g_\mathscr{B}\mu$ is a Borel measure*, the **Borel image measure*** under g. If f is a real valued Borel measurable function on $\mathbb{R}$ then $f \circ g$ is $\mathscr{L}\mathscr{A}$ measurable, and if g has a support E in $\mathscr{A}_\sigma$ and $f(0) = 0$, then $f \circ g$ also vanishes outside E. Consequently the preceding lemma applies, and $\int f \, dg_\mathscr{B}\mu = \int f \circ g \, d\mu$ if $f \in L_1(g_\mathscr{B}\mu)$, or if $f \circ g \in L_1(\mu)$. In particular, if f is the identity function $t \mapsto t$ for t in $\mathbb{R}$, then $\int g \, d\mu = \int t \, dg_\mathscr{B}\mu t$ for each g in $L_1(\mu)$.

3 Corollary *Suppose μ is a measure on $\mathscr{A}$, g is real valued and $\mathscr{A}$ σ-simple, and $g_\mathscr{B}\mu$ is the Borel image measure*. Then a Borel measurable real valued function f on $\mathbb{R}$ that vanishes at 0 is a member of $L_1(g_\mathscr{B}\mu)$ iff $f \circ g \in L_1(\mu)$, and in this case $\int f \, dg_\mathscr{B}\mu = \int f \circ g \, d\mu$.*

In particular, $\int g \, d\mu = \int t \, dg_\mathscr{B}\mu t$ for all g in $L_1(\mu)$.

It is worth noting that if g is μ integrable and $r > 0$ then $r\mu_\#(\{x: |g(x)| > r\}) \leqq \|g\|_1$. Consequently, the image $g_\mathscr{B}\mu$ is finite at each member of $\mathscr{B}^\delta(\mathbb{R})$ that is bounded away from 0, and if μ is not totally finite, each Borel neighborhood of 0 has infinite $g_\mathscr{B}\mu$ measure.

We will be concerned with mappings of Borel measures for intervals. The Borel δ-ring $\boldsymbol{\mathscr{B}^\delta(E)}$ of a (possibly infinite) interval E is the δ-ring generated by the family of compact subsets of E or, alternatively, by the closed intervals $[c:d]$ contained in E. If E is a closed interval $[a:b]$, then $\mathscr{B}^\delta[a:b]$ is the family $\mathscr{B}^\delta(\mathbb{R})\|[a:b]$ of all members of $\mathscr{B}^\delta(\mathbb{R})$ that are subsets of $[a:b]$. For each interval E, the Borel σ-field $\boldsymbol{\mathscr{B}(E)}$ is $\mathscr{L}\mathscr{B}^\delta(E)$, which is identical with $\mathscr{B}(\mathbb{R})\|E$. A **Borel measure* (measure) for E** is a measure* (measure, respectively) on $\mathscr{B}^\delta(E)$.

If E is a (possibly infinite) open interval, $E = (a:b)$ with $-\infty \leqq a < b \leqq \infty$, then there is a one-to-one monotonic continuous map with a continuous inverse on $(a:b)$ onto $\mathbb{R}$—for example, if a and b are finite, then $x \mapsto 1/(b - x) - 1/(x - a)$ for x in $(a:b)$, is such a map. Consequently results about Borel measures for $\mathbb{R}$ imply results about Borel measures for $(a:b)$. Here is an example.

If f is an increasing real valued function on $(a:b)$, then the "length function" given by $[c:d] \mapsto f_+(d) - f_-(c)$ for each closed subinterval $[c:d]$ of $(a:b)$, is a pre-measure (see chapter 2) and hence, by proposition 4.3, it extends to a unique Borel measure ν_f for $(a:b)$. The measure ν_f is the **Borel measure for** $(a:b)$ **induced by** f. If g is another increasing function on $(a:b)$ then $\nu_g = \nu_f$ iff g is sandwiched between $c + f_-$ and $c + f_+$ for some constant c. Every Borel measure μ for $(a:b)$ is induced by some f—for example (proposition 1.4), if $c \in (a:b)$, by $x \mapsto \mu[c:x] - \mu\{c\}$ for x in $[c:b)$ and $x \mapsto -\mu[x:c] + \mu\{x\}$ for x in $(a:c)$. Each increasing function inducing μ is called a **distribution function** for μ. Such a function normalized to be right continuous, is unique to an additive constant.

Borel–Lebesgue measure $\Lambda_{(a:b)}$ for $(a:b)$ is just the restriction of Borel–Lebesgue measure Λ to the δ-ring $\mathscr{B}^\delta(a:b)$. We shall be concerned with the Borel image $F\Lambda_{(a:b)}$ of $\Lambda_{(a:b)}$ under an increasing real valued function F on $(a:b)$. Such a function F is $\mathscr{B}(a:b) - \mathscr{B}(\mathbb{R})$ measurable and is continuous except at the points of a countable set. In particular, F differs from the right continuous function F_+ only at the points of a countable set, consequently the symmetric difference of $\{x: F(x) \in E\}$ and $\{x: F_+(x) \in E\}$ has Λ measure zero for all E, and hence $F\Lambda_{(a:b)} = F_+\Lambda_{(a:b)}$. In a similar fashion one sees that $f\Lambda_{(a:b)}$ is the same for all f sandwiched between F_- and F_+.

The **quasi-inverse** $F^\sim$ of an increasing function F on $(a:b)$ to $\mathbb{R}$ is defined as follows. Let $\alpha = \mathit{inf}_x F(x)$ and $\beta = \mathit{sup}_x F(x)$. For each member t of the open interval $(\alpha:\beta)$ the set $\{x: F(x) \geqq t\}$ is not empty and is bounded below and we define $F^\sim(t)$ to be $\mathit{inf}\{x: F(x) \geqq t\}$ for t in $(\alpha:\beta)$. Evidently $F^\sim$ is just F^{-1} if F is continuous and one-to-one. We note that $F^\sim = (F_+)^\sim$, and that $F^\sim$ is always left continuous (if $\{t_n\}_n$ is an increasing sequence in $(\alpha:\beta)$ converging to a member t of $(\alpha:\beta)$ then $\{\{x: F_+(x) \geqq t_n\}\}_n$ is a decreasing sequence of half infinite intervals whose intersection is $\{x: F_+(x) \geqq t\}$, whence $\mathit{lim}_n F_+{}^\sim(t_n) = F_+{}^\sim(t)$).

4 THEOREM *Suppose F is an increasing real valued function on a (possibly infinite) interval $(a:b)$, $\alpha = \inf_x F(x)$ and $\beta = \sup_x F(x)$.*

Then the Borel image $F\Lambda_{(a:b)}$, is the Borel measure induced by $F^\sim$, and the Borel measure induced by F is the Borel image measure $F^\sim\Lambda_{(\alpha:\beta)}$.

PROOF We may assume without loss of generality that F is right continuous, and hence that $F^\sim(t)$ is the smallest member of $\{x: F(x) \geqq t\}$ for each t in $(\alpha:\beta)$. Consequently $F(x) < t$ iff $x < F^\sim(t)$, whence $F(x) \geqq t$ iff $x \geqq F^\sim(t)$.

If $\alpha < u < v < \beta$, then $\Lambda_{(a:b)}(F^{-1}[u:v)) = \Lambda(\{x: u \leqq F(x) < v\}) =$

$\Lambda_{(a:b)}(\{x\colon F^\sim(u) \leqq x < F^\sim(v)\}) = F^\sim(v) - F^\sim(u)$. Since $F^\sim$ is left continuous, $F^\sim(v) - F^\sim(u) = v[u:v)$ where v is the measure induced by F. It follows that $F\Lambda_{(a:b)}$ agrees with v on $\mathscr{B}^\delta(\alpha:\beta)$.

If $a < c < d < b$, then $F^\sim\Lambda_{(\alpha:\beta)}(c:d] = \Lambda_{(\alpha:\beta)}(\{t\colon c < F^\sim(t) \leqq d\}) = \Lambda_{(\alpha:\beta)}(\{t\colon F(c) < t \leqq F(d)\}) = F(d) - F(c)$. Hence $F^\sim\Lambda_{(\alpha:\beta)}$ is the measure induced by F. ∎

A particular consequence of the preceding result is that each Borel measure μ for an open interval is the Borel image of $\Lambda_{(\alpha:\beta)}$ for some α and β, because μ is induced by an increasing function—any distribution function F for μ will do. We obtain, after a preliminary lemma, a representation theorem for μ in the case that F is continuously differentiable.

Suppose v is a measure on a δ-ring $\mathscr{A}$ of subsets of X and f is **locally v integrable** on X, in the sense that $f\chi_A \in L_1(v)$ for each A in $\mathscr{A}$. In this case, the **indefinite integral $f.v$** of f w.r.t. v is defined by **$f.v(A)$** $= \int f\chi_A\,dv$ for A in $\mathscr{A}$. (We shall later extend the notion of indefinite integral to a more general situation.)

5 LEMMA *If w is non-negative and locally v integrable, then the indefinite integral $w.v$ is a measure on $\mathscr{A}$, $f \in L_1(w.v)$ iff $fw \in L_1(v)$, and in this case $\int f\,d(w.v) = \int fw\,dv$.*

PROOF If $\{A_n\}_n$ is a disjoint sequence in $\mathscr{A}$ and $A = \sum_n A_n \in \mathscr{A}$, then $lim_N \sum_{n=1}^N w.v(A_n) = \int w\chi_A\,dv = w.v(A)$ by B. Levi's theorem and it follows that $w.v$ is a measure on $\mathscr{A}$. If f is a non-negative member of $L_1(w.v)$, then $f = \sum_n b_n\chi_{B_n}$ for some $b_n \geqq 0$ and B_n in $\mathscr{A}$ for each n, so $fw = \sum_n b_n w\chi_{B_n}$. Hence $\int fw\,dv = \sum_n b_n \int w\chi_{B_n}\,dv = \sum_n b_n w.v(B_n) = \int f\,d(w.v)$. ∎

If a distribution function F of a Borel measure μ for a (possibly infinite) interval $(a:b)$ is continuously differentiable, then $\mu[c:d] = \int_c^d F'(t)\,dt = \int F'\chi_{[c:d]}\,d\Lambda_{(a:b)} = F'.\Lambda_{(a:b)}[c:d]$ for each closed subinterval $[c:d]$ of $(a:b)$. Consequently, since $F'.\Lambda_{(a:b)}$ is a measure that agrees with μ on closed intervals, it agrees with μ on $\mathscr{B}^\delta(a:b)$.

6 PROPOSITION *If a distribution function F of a Borel measure μ for $(a:b)$ is continuously differentiable, then $\mu = F'.\Lambda_{(a:b)}$, and hence $\int f\,d\mu = \int fF'\,d\Lambda_{(a:b)}$ for all f in $L_1(\mu)$.*

The following result is an easy consequence of proposition 6 and theorem 4, which give two different descriptions of the measure induced by a continuously differentiable distribution function. However, a direct proof is easy as well. (An extension of the result is given in a supplement.)

7 Proposition (Images of $\Lambda_{(a:b)}$ Under Smooth Maps) *If T is a continuously differentiable map of $(a:b)$ onto $(\alpha:\beta)$ and T' is never zero, then the Borel image $T_{\mathscr{B}}\Lambda_{(a:b)}$ is the indefinite integral $|(T^{-1})'|.\Lambda_{(\alpha:\beta)}$.*

PROOF If $[\gamma:\delta] \subset (\alpha:\beta)$, then $T_{\mathscr{B}}\Lambda_{(a:b)}[\gamma:\delta] = \Lambda_{(a:b)}(T^{-1}[\gamma:\delta])$. If $T'(x) > 0$ for each x, then $\Lambda_{(a:b)}(T^{-1}[\gamma:\delta])$ is $T^{-1}(\delta) - T^{-1}(\gamma) = \int_{[\gamma:\delta]}(T^{-1})'\,d\Lambda_{(\alpha:\beta)} = (T^{-1})'.\Lambda_{(\alpha:\beta)}[\gamma:\delta]$, and if $T'(x) < 0$ for each x, then $\Lambda_{(a:b)}T^{-1}[\gamma:\delta]$ is $T^{-1}(\gamma) - T^{-1}(\delta) = \int_{[\gamma:\delta]} -(T^{-1})'\,d\Lambda_{(\alpha:\beta)} = |(T^{-1})'|.\Lambda_{(\alpha:\beta)}[\gamma:\delta]$. The proposition follows. ■

It is worth noticing that the usual convention for a definite integral disguises the necessary absolute value in the statement of the preceding result. Explicitly: suppose T is continuously differentiable on $(a:b)$ onto $(\alpha:\beta)$, $T'(x) > 0$ for all x, $[c:d] \subset (a:b)$, and g is a non-negative Borel measurable function on $(\alpha:\beta)$. Then, according to the preceding proposition and the mapping lemma 2, $\int(\chi_{(c:d)}g)\circ T\,d\Lambda_{(a:b)} = \int(\chi_{(c:d)}g)(T^{-1})'\,d\Lambda_{(\alpha:\beta)}$. In the usual notation for definite integrals, this becomes $\int_{T^{-1}(c)}^{T^{-1}(d)} g(T(x))\,d\Lambda_{(a:b)}x = \int_c^d g(y)(T^{-1})'(y)\,d\Lambda_{(\alpha:\beta)}y$. This formula is also correct if T' is always negative, *provided* we agree that $\int_{T^{-1}(c)}^{T^{-1}(d)}$ is $-\int_{T^{-1}(d)}^{T^{-1}(c)}$ if $T^{-1}(d) < T^{-1}(c)$.

A variant of the formula $\int_{T^{-1}(c)}^{T^{-1}(d)} g\circ T\,d\Lambda_{(a:b)} = \int_c^d g/T'\,d\Lambda_{(\alpha:\beta)}$ is obtained by setting $g = f\circ S$ and $S = T^{-1}$. We then have $\int_{S(c)}^{S(d)} f\,d\Lambda_{(a:b)} = \int_c^d f(S(y))S'(y)\,d\Lambda_{(\alpha:\beta)}y$ for each non-negative Borel measurable function f on $(a:b)$ and each continuously differentiable map S of $(\alpha:\beta)$ onto $(a:b)$ such that S' is everywhere positive. This is sometimes called the "**change of variable**" formula. If f is continuous and hence the derivative of some function F on $(a:b)$, then the proposition becomes $\int_{S(c)}^{S(d)} F'\,d\Lambda_{(a:b)} = \int_c^d (F\circ S)'\,d\Lambda_{(\alpha:\beta)}$, and in this case the formula follows from the fundamental theorem of calculus. We observe: this change of variable formula does not require that S' be non-vanishing nor that S be one-to-one.

It is important to characterize Borel measures μ that are indefinite integrals w.r.t. Λ. Each such measure vanishes at each Borel set of Λ measure zero, but it would be a bold conjecture that each Borel measure μ vanishing at Λ null Borel sets is an indefinite integral with respect to Λ. This is in fact the case—it is a special case of the Radon–Nikodym theorem established in the next chapter. Here we content ourselves with showing how to recover the function w (at least Λ a.e.) from the measure $w.\Lambda$.

For each locally Λ integrable function w and for $h > 0$, let $\boldsymbol{A_h(w)}(x) = (1/h)\int_x^{x+h} w\,d\Lambda = (1/h)w.\Lambda((x:x+h))$ and let $T_t(w)(x) = w(x+t)$. Thus $A_h(w)(x)$ is the average w.r.t. Λ of w over $(x:x+h)$.

8 Lemma *If $w \in L_1(\Lambda)$ and $h > 0$, then $\|A_h(w) \dot{-} w\|_1 \leqq (1/h)\int_{s=0}^h \|T_s(w) - w\|_1\,d\Lambda s \leqq \sup\{\|T_s(w) - w\|_1 : s \in [0:h]\}$.*

PROOF We see: $\|A_h(w) - w\|_1 = \int |(1/h)\int_{t=x}^{x+h}((w(t) - w(x))\,d\Lambda t|\,d\Lambda x$ $\leqq (1/h)\int\int_{(x:x+h)}|w(t) - w(x)|\,d\Lambda t\,d\Lambda x$ which, because Λ is translation invariant, is $(1/h)\int\int_{[0:h]}|w(x+t) - w(x)|\,d\Lambda t\,d\Lambda x$, which by the Fubini theorem is $(1/h)\int_{[0:h]}\int|T_s(w)(x) - w(x)|\,d\Lambda x\,d\Lambda s$. This establishes the first inequality and since $\|T_s(w) - w\|_1 \leqq 2\|w\|_1 < \infty$, the second is clear. ∎

The preceding lemma shows that if $w \in L_1(\Lambda)$, then $A_h(w)$ converges in the norm of $L_1(\Lambda)$ to w as h goes to 0, since $s \mapsto T_s(w)$ is a continuous function on $\mathbb{R}$ to $L_1(\Lambda)$ according to 6.11 and $T_0(w) = w$. If $\mu = w.\Lambda$, then $A_h(w)(x)$ is $\mu((x: x+h))/\Lambda((x: x+h))$, which is sometimes abbreviated as $(\Delta_h\mu/\Delta_h\Lambda)(x)$, and its limit w, as h goes to 0, is denoted $(d\mu/d\Lambda)(x)$ (it is called a **Radon–Nikodym derivative** of μ w.r.t. Λ and is determined Λ a.e. only). If F is the distribution function of μ, then $A_h(w)$ at x is the difference quotient $(F(x+h) - F(x))/h$. We record:

9 THEOREM (LEBESGUE) *If $w \in L_1(\Lambda)$ and $\mu = w.\Lambda$, then $\Delta_h\mu/\Delta_h\Lambda$ belongs to $L_1(\Lambda)$ and converges in norm to w as h goes to zero.*

If w is an arbitrary locally Λ integrable function, then the measure $w.\Lambda$ still determines w, Λ a.e., because the preceding applies to $w\chi_{[a:b]}$ for each interval $[a:b]$.

If F is a distribution function of $w.\Lambda$, the preceding version of Lebesgue's theorem yields that $(1/h)(F(x+h) - F(x))$ converges in L_1-norm to w as h goes to zero. The best version of Lebesgue's theorem asserts the pointwise convergence of $(1/h)(F(x+h) - F(x))$ to $w(x)$ for almost every x. However, the pointwise analysis takes additional machinery. A good account of these matters can be found in (E. Hewitt and K. Stromberg [1], I. E. Segal and R. A. Kunze [1], and R. L. Wheeden and A. Zygmund [1].)

10 NOTES

(i) A Borel measure which is an indefinite integral w.r.t. Λ vanishes at each singleton $\{x\}$, so the corresponding distribution function F is continuous, and moreover, F is the indefinite integral of its derivative (which will exist a.e. by Lebesgue's theorem). On the other hand there are continuous (and even continuous and strictly increasing) distribution functions F with $F'(x) = 0$ for almost every x—the corresponding Borel measures are *singular* and are carried on sets in $\mathscr{B}(\mathbb{R})$ which are Λ null (see chapter 9).

(ii) Oxtoby and Ulam have characterized Borel measures μ which are Borel images of Λ under a homeomorphism of $\mathbb{R}$ onto itself—μ is such a measure iff $\mu\{x\} = 0$ for each x, $\mu(U) > 0$ for each open Borel set U and μ is not totally finite. See *Ann. Math.* **42** (1941), 874–920.

SUPPLEMENT: STIELTJES INTEGRATION

We have seen (Corollary 3) that an arbitrary integral $\int g\,d\mu$ can be computed as an integral with respect to a Borel measure $v = g_{\mathscr{B}}\mu$ for $\mathbb{R}$. We will show that, at least for a continuous function f on a closed interval $[a:b]$, $\int f\,d\mu$ can be computed conveniently by a Riemann type approximation in terms of a distribution function for μ; that is, as a Riemann–Stieltjes integral.

Suppose that F is an increasing function on the closed interval $[a:b]$ and that $-\infty < a < b < \infty$. Then the extension of F to $\mathbb{R}$ that is constant on $(-\infty:a]$ and on $[b:\infty)$ induces a measure on $\mathscr{B}^\delta(\mathbb{R})$. This measure, restricted to the δ-ring $\mathscr{B}^\delta[a:b]$ of those members of $\mathscr{B}^\delta(\mathbb{R})$ that are subsets of $[a:b]$, is a Borel measure for $[a:b]$—it is the **measure induced by *F***. If μ is the measure induced by F then $\mu\{a\} = F_+(a) - F(a)$, $\mu\{b\} = F(b) - F_-(b)$ and $\mu[c:d] = F_+(d) - F_-(c)$ for all c and d in $(a:b)$.

Each Borel measure for $[a:b]$ is induced by an increasing function on $[a:b]$ just as each Borel measure for $\mathbb{R}$ is induced by a distribution function on $\mathbb{R}$. We see this as follows. A Borel measure μ for $[a:b]$ has a natural extension $\mu^\wedge$ to a Borel measure for $\mathbb{R}$, given by $\mu^\wedge(A) = \mu(A \cap [a:b])$ for A in $\mathscr{B}^\delta(\mathbb{R})$. Every distribution function for $\mu^\wedge$ is constant on $(-\infty:a)$ and on $(b:\infty)$, and there are such distribution functions that are constant on $(-\infty:a]$ and on $[b:\infty)$. The restriction of one of these to $[a:b]$ induces μ.

Suppose that μ is the Borel measure for $[a:b]$ induced by an increasing real valued function F on $[a:b]$. We describe a method of computing $\int f\,d\mu$—at least for suitable functions f—directly in terms of the distribution function F.

A **subdivision** $\sigma = \{\sigma_i\}_{i=1}^{q+1}$ of $[a:b]$ is a finite sequence σ such that $a = \sigma_1 < \sigma_2 < \cdots < \sigma_{q+1} = b$. For a bounded real valued function f on $[a:b]$ let m_i and M_i be the infimum and supremum respectively of f on the open interval $(\sigma_i:\sigma_{i+1})$. We adopt, for convenience, the convention that $F_-(\sigma_1) = F(\sigma_1)$ and $F_+(\sigma_{q+1}) = F(\sigma_{q+1})$, and we set $\boldsymbol{L(\sigma,f,F)} = \sum_{i=1}^{q+1} f(\sigma_i)(F_+(\sigma_i) - F_-(\sigma_i)) + \sum_{i=1}^{q} m_i(F_-(\sigma_{i+1}) - F_+(\sigma_i))$ and $\boldsymbol{U(\sigma,f,F)}$ the same sum with m_i replaced by M_i. If v is the step function that agrees with f at points of the subdivision and is m_i on each interval $(\sigma_i:\sigma_{i+1})$, then $v \leqq f$ and $L(\sigma,f,F) = \int v\,d\mu$, where μ is the measure induced by F. Similarly, there is a step function u, $f \leqq u$ such that $U(\sigma,f,F) = \int u\,d\mu$. It follows that if f is μ integrable, then $L(\sigma,f,F) \leqq \int f\,d\mu \leqq U(\sigma,f,F)$.

The function f is **Stieltjes integrable**, or **Riemann–Stieltjes integrable** w.r.t. F over $[a:b]$ iff the supremum, $sup_\sigma L(\sigma,f,F)$, over all subdivisions σ of $[a:b]$ is equal to $inf_\sigma U(\sigma,f,F)$, and in this case the **Stieltjes integral** $\boldsymbol{\int_a^b f(t)\,dF(t)}$ is $sup_\sigma L(\sigma,f,F) = inf_\sigma U(\sigma,f,F)$. Evidently: if f is μ integrable and is also Stieltjes integrable w.r.t. F, then $\int f\,d\mu = \int f(t)\,dF(t)$.

Finally, each continuous function f is Stieltjes integrable w.r.t. F. This follows at once from the uniform continuity of f and the observation that $U(\sigma, f, F) - L(\sigma, f, F) = \sum_{i=1}^{q} (M_i - m_i)(F_-(\sigma_{i+1}) - F_+(\sigma_i))$. Thus:

11 PROPOSITION *If a bounded function f on $[a:b]$ is Stieltjes integrable w.r.t. an increasing function F and is also integrable w.r.t. the measure μ induced by F—and in particular if f is continuous—then $\int_a^b f(t)\,dF(t) = \int f\,d\mu$.*

If $F(t) = t$ for each t in $[a:b]$, then the Stieltjes integral $\int_a^b f(t)\,dF(t)$ is just the Riemann integral $\int_a^b f(t)\,dt$. We notice that, with our convention, $\int_a^b g\,dP = \int_a^b g(t)\,dPt$ would denote the integral of g w.r.t. a measure P, whereas $\int_a^b g(t)\,dP(t)$ denotes the Stieltjes integral w.r.t. a function P on $[a:b]$.

NOTE A bounded function f on $[a:b]$ may be Stieltjes integrable w.r.t. an increasing function F without being continuous on $[a:b]$—for instance $f = \chi_{\{a\}}$. It turns out (see H. J. TerHorst, *Amer. Math. Monthly* **91** (1984), 551–559) that f is Stieltjes integrable w.r.t. F iff f is continuous μ a.e. on the set $Z_\mu = \{x \in [a:b] : \mu\{x\} = 0\}$. In particular, f is Riemann integrable iff the set of points of discontinuity has Lebesgue measure zero. ■

It is not surprising that some standard theorems involving Borel measures were originally results about Stieltjes integration. The Riesz representation theorem 6.5 is a classical instance. Here is (approximately) the original form of the result.

12 COROLLARY *Suppose I is a positive linear functional on the space C of real valued continuous functions on $[0:1]$, and $F(t) = \inf\{I(h): h \in C$ and $h \geqq \chi_{[0:t]}\}$ for $0 < t \leqq 1$ and $F(0) = 0$.*
Then $I(f) = \int_0^1 f(t)\,dF(t)$ for all f in C.

PROOF The functional $f \mapsto I(f \mid [0:1])$, for f in the class $C_c(\mathbb{R})$ of continuous functions on $\mathbb{R}$ with compact support, is a positive linear functional on $C_c(\mathbb{R})$. Hence, by 6.5, there is a Borel measure ν for $\mathbb{R}$ such that $I(f \mid [0:1]) = \int f\,d\nu$ for f in $C_c(\mathbb{R})$. The measure ν vanishes on Borel sets outside $[0:1]$ because: if $[a:b]$ is disjoint from $[0:1]$, then there is f in $C_c(\mathbb{R})$ which is 1 on $[a:b]$, zero on $[0:1]$ and $0 \leqq f \leqq 1$, whence $\nu[a:b] \leqq \int f\,d\nu = I(f \mid [0:1]) = 0$. It is straightforward to verify that F induces $\nu \| \mathscr{B}^\delta[0:1]$, and the corollary follows. ■

A Stieltjes integral $\int_a^b f(t)\,dG(t)$ reduces to a Riemann integral if G is an increasing function that is sufficiently smooth. In particular this is the case if G has a Riemann integrable derivative.

13 PROPOSITION *Suppose g is a non-negative Riemann integrable function on $[a:b]$ and $G(x) = \int_a^x g(t)\,dt + c$ for x in $[a:b]$. Then each Riemann integrable function f is Stieltjes integrable over $[a:b]$* w.r.t. G *and $\int_a^b f(t)\,dG(t) = \int_a^b f(t)g(t)\,dt$.*

PROOF Suppose $\sigma = \{\sigma_i\}_{i=1}^{q+1}$ is a subdivision of $[a:b]$, M_i and m_i are respectively the supremum and infimum of f on $(\sigma_i:\sigma_{i+1})$ for each i and M is the supremum of g on $[a:b]$. Since G is continuous the upper Riemann–Stieltjes sum $U(\sigma, f, G)$ reduces to $\sum_{i=1}^q M_i(G(\sigma_{i+1}) - G(\sigma_i))$ whence $U(\sigma, f, G) - \int_a^b f(t)g(t)\,dt = \sum_{i=1}^q \int_{\sigma_i}^{\sigma_{i+1}} (M_i - f(t))g(t)\,dt$. $L(\sigma, f, G) - \int_a^b f(t)g(t)\,dt = \sum_{i=1}^q \int_{\sigma_i}^{\sigma_{i+1}} (m_i - f(t))g(t)\,dt$. Thus $U(\sigma, f, G) - L(\sigma, f, G)$ is majorized by $M\sum_{i=1}^q \int_{\sigma_i}^{\sigma_{i+1}} (M_i - m_i)\,dt = M\sum_{i=1}^q (M_i - m_i)(\sigma_{i+1} - \sigma_i)$, which is small if $max_i\, \sigma_i$ is small, because f is Riemann integrable. The proposition follows. ■

SUPPLEMENT: THE IMAGE OF Λ_p UNDER A SMOOTH MAP

Proposition 7 on the image of $\Lambda_{(a:b)}$ under a smooth map T generalizes to higher dimensions. We consider first a linear map $T: \mathbb{R}^p \to \mathbb{R}^p$. The determinant of T is denoted **det** $\boldsymbol{T}$.

14 THEOREM *If a linear map T: $\mathbb{R}^p \to \mathbb{R}^p$ is non-singular and $f \in C_c(\mathbb{R}^p)$ then $\int f\,d\Lambda_p = \int |\det T|\, f \circ T\,d\Lambda_p$.*

PROOF According to proposition 4.14 the integral of a member f of $C_c(\mathbb{R}^p)$ w.r.t. Λ_p can be obtained by iterated integration w.r.t. Λ. It follows that if R is a simple reflection in the plane $x_i = x_j$, $i \neq j$, so $R(x)_i = x_j$, $R(x)_j = x_i$ and $R(x)_k = x_k$ for k different from i and j, then $\int f\,d\Lambda_p = \int f \circ R\,d\Lambda_p$. Evidently $|det\, R| = 1$. More generally, if R is a composition of such simple reflections (for varying i and j), then $f\,d\Lambda_p = \int |det\, R|\, f \circ R\,d\Lambda_p$ and $|det\, R| = 1$.

A linear map S: $\mathbb{R}^p \to \mathbb{R}^p$ is a simple shear if for some i, j with $i \neq j$ and some scalar t, $S(x)_i = x_i + tx_j$ and $S(x)_k = x_k$ for $k \neq i$ and for all x. Proposition 4.14 shows that for all simple shears, and hence for each composition S of simple shears, $\int f\,d\Lambda_p = \int f \circ S\,d\Lambda_p$ for all f in $C_c(\mathbb{R}^p)$. Evidently $|det\, S| = 1$ for each such composition.

Proposition 4.14 also shows that if D is a non-singular diagonal map of $\mathbb{R}^p$ to $\mathbb{R}^p$ (that is, if for some nonzero $\lambda_1, \lambda_2 \ldots \lambda_p$ and for all x, $D(x)_i = \lambda_i x_i$), then $\int f\,d\Lambda_p = \int |\lambda_1, \lambda_2 \ldots \lambda_p|\, f \circ D\,d\Lambda_p = \int |det\, D|\, f \circ D\,d\Lambda_p$.

Finally: A square matrix can be reduced to a diagonal matrix by elementary row operations, and it follows that if T is an arbitrary linear map, then $S \circ T \circ R$ is a diagonal map D for some composition S of simple shears and some composition R of simple reflections. But the inverse of a simple shear is a simple shear and each reflection is its own inverse and so T is of the form $S \circ D \circ R$. Hence $|det\, T| = |det\, S\, det\, D\, det\, R| = |det\, D|$ and so $\int |det\, T|\; f \circ T\, d\Lambda_p = \int |det\, S \circ D \circ R| f \circ S \circ D \circ R\, d\Lambda_p = \int |det\, D|\; f \circ S \circ D\, d\Lambda_p = \int f \circ S\, d\Lambda_p = \int f\, d\Lambda_p$. ∎

A function T on an open subset of $\mathbb{R}^p$ to $\mathbb{R}^p$ is **differentiable** at a point x of its domain and $T'(x)$ is its **derivative**, iff $T'(x)$ is a linear map of $\mathbb{R}^p$ to $\mathbb{R}^p$ such that $(T(x+h) - T(x) - T'(x)(h))/\|h\|$ converges to zero as $\|h\|$ converges to zero. The matrix of $T'(x)$ with respect to the usual basis for $\mathbb{R}^p$, which has $i-j$ entry $\partial(T(x))_i/\partial x_j$, is called the **Jacobi matrix** at x and the absolute value of its determinant, which is the same as $|det\, T'(x)|$, is the **Jacobian of T at** x.

A map T: $\mathbb{R}^p \to \mathbb{R}^p$ is called **affine** iff it differs from a linear transformation by a constant; that is, $T(x) = L(x) + c$ for some linear map L, some c and all x in $\mathbb{R}^p$. In this case $T'(x)$, for every x, is the linear map L, and in fact affine maps are just those that have constant derivatives.

If L: $\mathbb{R}^p \to \mathbb{R}^p$ is linear and $C(x) = x + c$ is translation by c, then $C \circ L(x) = L(x) + c = T(x)$ and since integration w.r.t. Λ_p is invariant under translation, the preceding theorem implies that $\int f\, d\Lambda_p = \int |det\, T'| f \circ T\, d\Lambda_p$ for all f in $C_c(\mathbb{R}^p)$, provided $|det\, T'| \neq 0$. We agree that an affine map T is **non-singular** if $|det\, T'| \neq 0$.

15 COROLLARY *If* T: $\mathbb{R}^p \to \mathbb{R}^p$ *is a non-singular affine map, then* $\int f\, d\Lambda_p = \int |\det T'| f \circ T\, d\Lambda_p$ *for all* f *in* $C_c(\mathbb{R}^p)$, *and* $\Lambda_p(T[A]) = |\det T'|\, \Lambda_p(A)$ *for all* A *in* $\mathscr{B}^\delta(\mathbb{R}^p)$.

PROOF We prove the second assertion. If B is a compact subset of $\mathbb{R}^p$, then there is a decreasing sequence $\{f_n\}_n$ in $C_c(\mathbb{R}^p)$ that converges pointwise to χ_B. Consequently $\Lambda_p(B) = lim_n \int f_n\, d\Lambda_p = lim_n |det\, T'| \times \int f_n \circ T\, d\Lambda_p = |det\, T'| \int \chi_B \circ T\, d\Lambda_p = |det\, T'|\, \Lambda_p(T^{-1}[B])$, and if $B = T[A]$, then $\Lambda_p(T[A]) = |det\, T'|\, \Lambda_p(A)$. Each open member of $\mathscr{B}^\delta(\mathbb{R}^p)$ is the union of an increasing sequence of compact sets A, and it follows that $\Lambda_p(T[U] = |det\, A'|\, \Lambda_p(U)$. The same equality holds for an arbitrary member B of $\mathscr{B}^\delta(\mathbb{R}^p)$ because of regularity. ∎

Each differentiable function T on $\mathbb{R}^p$ to $\mathbb{R}^p$ is, intuitively, "locally almost affine," and this suggests that the preceding result, together with a subdivision argument, could show that $\Lambda_p(T[A]) = \int_A |det\, T'|\, d\Lambda_p$ for suitable T and A. We establish a theorem of this sort after a couple of lemmas.

The **half open cube $Q_r(a)$** in $\mathbb{R}^p$ with center $a = (a_1, a_2 \ldots a_p)$ and half width $r > 0$ is the cartesian product $\times_{i=1}^p [a_i - r : a_i + r)$. Norm $\mathbb{R}^p$ by setting $\|x\|_\infty = sup\{|x_i|: i = 1, 2 \ldots p\}$. Then the open ball about a of radius r is contained in $Q_r(a)$, $Q_r(a)$ is contained in the closed ball, and its diameter is $2r$. As $Q_r(a)$ can be subdivided by bisection into 2^p

disjoint half-open cubes of half width $r/2$, it follows that each half open cube can be subdivided into disjoint half open cubes of small diameter. Further:

16 LEMMA *Each open subset W of $\mathbb{R}^p$ is the union of a disjoint countable family of half open cubes Q of small diameter whose closures are subsets of W.*

PROOF Let $\mathscr{A}$ be a disjoint countable family of half open cubes that covers $\mathbb{R}^p$. For $e > 0$ let $\mathscr{B}$ be the family of all half open cubes Q that are obtained from members of $\mathscr{A}$ by successive bisections and for which $Q^- \subset W$ and $dia(Q) < e$, and let $\mathscr{C}$ be the subfamily of $\mathscr{B}$ consisting of those members of $\mathscr{B}$ that are proper subsets of no member of $\mathscr{B}$. Evidently $\mathscr{C}$ is a disjoint family, and for each x in W the class $\{Q: x \in Q \in \mathscr{B}\}$ is non-empty and linearly ordered by $\subset$ and its largest member belongs to $\mathscr{C}$, so $\mathscr{C}$ covers W. ■

The space of all linear maps of $\mathbb{R}^p$ to $\mathbb{R}^p$ is normed by setting $\|A\|$, the **operator norm** of A, to be $sup\{\|A(x)\|_\infty: x \in \mathbb{R}^p \text{ and } \|x\|_\infty \leqq 1\}$ for each map A, and so $\|A(y)\|_\infty \leqq \|A\| \, \|y\|_\infty$ for all y in $\mathbb{R}^p$. In particular, we note that if I is the identity map of $\mathbb{R}^p$, then $A(h) = h + (A - I)(h) = h + (\textit{a vector of length at most } \|A - I\| \, \|h\|_\infty)$.

A map T on an open subset U of $\mathbb{R}^p$ to $\mathbb{R}^p$ is **continuously differentiable** iff $x \mapsto T'(x)$ is a continuous map on U to the space of all linear maps of $\mathbb{R}^p$ to $\mathbb{R}^p$ with the operator norm. If T is continuously differentiable and one to one on U, and the Jacobian of T does not vanish, then by the inverse function theorem, $V = T[U]$ is open, T^{-1} is continuously differentiable on V and $(T^{-1})'(Tx) = (T'(x))^{-1}$ for x in U.

For each open subset U of $\mathbb{R}^p$ let $\boldsymbol{\mathscr{B}^\delta(U)}$ be the δ-ring generated by the family of compact subsets of U, or what is the same thing, the class of members B of $\mathscr{B}^\delta(\mathbb{R}^p)$ such that the closure B^- of B is a subset of U. Let $\boldsymbol{\Lambda_U}$ be Λ_p restricted to the family $\mathscr{B}^\delta(U)$.

17 LEMMA *Suppose $T: U \to \mathbb{R}^p$ is one to one, continuously differentiable, and has a non-vanishing Jacobian on an open subset U of $\mathbb{R}^p$, that $V = T[U]$ and that $B \in \mathscr{B}^\delta(U)$. Then $\inf\{|det\, T'(c)|: c \in B\}\Lambda_U(B) \leqq \Lambda_V(T[B]) \leqq \sup\{|det\, T'(c): c \in B\}\Lambda_U(B)$.*

PROOF Suppose first that W is a bounded open set with $W^- \subset U$ and a and $a + h$ are members of $\mathbb{R}^p$ such that the line segment $\{a + th: 0 \leqq t \leqq 1\}$ is contained in W. Then, because $(d/dt)(T(a + th) = T'(a + th)(h)$, $T(a + h) - T(a) = \int_{t=0}^1 T'(a + th)(h)\, dt$, where the Riemann integral of an integrable $\mathbb{R}^p$ valued function F on $[0:1]$ is given by $(\int_{t=0}^1 F(t)\, dt)_i = \int_{t=0}^1 F(t)_i\, dt$ for $i = 1, 2, \ldots p$. Evidently $\|\int_{t=0}^1 F(t)\, dt\|_\infty \leqq max_i\, sup_t\, |F(t)_i|$.

If $c \in U$, $T'(c)^{-1}(T(a+h) - T(a)) = \int_{t=0}^{1} T'(c)^{-1} \circ T'(a+th)(h)\,dt = h + \int_{t=0}^{1}(T'(c)^{-1} \circ T'(a+th) - I)(h)\,dt$. If c, a and $a+h$ all belong to a half-open cube Q and $Q^- \subset W$, then $T'(c)^{-1}(T(a+h) - T(a)) = h +$ (*a vector of length at most* $\sup\{\|T'(c)^{-1} \circ T'(b) - I\|: b \in Q$ *and* $c \in Q\}\,\|h\|_\infty$). But T' is continuous and invertible at each point of the compact set Q^-, and a uniform continuity argument yields for $e > 0$ a $d > 0$ such that $sup\{\|T'(c)^{-1} \circ T'(b) - I\|: b \in Q$ and $c \in Q\} < e$ if $dia\,Q < d$. According to the preceding lemma, W is the union of a disjoint countable family $\mathscr{C}$ of cubes of diameter less than d. If Q is one of these with center a and width w, then $0 < w \leqq d$ and $T'(c)^{-1}(T(a+h) - T(a)) = h +$ (*a vector of length at most* $e\|h\|_\infty$), whence $T'(c)^{-1}[T[Q] - T(a)] \subset \{x: \|x\|_p < w(1+e)\}$, so $\Lambda_p(T'(c)^{-1}[T[Q]]) \leqq (1+e)^p \Lambda_U(Q)$. Corollary 15 then implies that $|det\,T'(c)^{-1}|\,\Lambda_V(T([Q]) \leqq (1+e)^p \Lambda_U(Q)$ for each c in Q. Consequently $\Lambda_V(T[Q]) \leqq sup\{|\det T'(c)|: c \in W\}(1+e)^p \Lambda_U(Q)$, and summing for Q in $\mathscr{C}$ establishes the same inequality for W, and for all $e > 0$. Hence $\Lambda_V(T[W]) \leqq sup\{|\det T'(c)|: c \in W\}\Lambda_U(W)$ for all open sets W with $W^- \subset U$.

Suppose that $B \in \mathscr{B}^\delta(U)$, $e > 0$ and W is a bounded open set such that $B \subset W$, $W^- \subset U$ and $\Lambda_U(W) < \Lambda_U(B) + e$. Let $b = sup\{|\det T'(c)|: c \in B\}$ and $W_e = \{x \in W: |det\,T'(x)| < b + e\}$. Then $B \subset W_e \subset W$ and by the preceding result, $\Lambda_V(T[B]) \leqq \Lambda_V(T[W_e]) \leqq (b+e)\Lambda_U(W_e) \leqq (b+e)(\Lambda_U(B) + e)$. Consequently $\Lambda_V(T[B]) \leqq b\Lambda_U(B) = sup\{|\det T'(c)|: c \in B\}\Lambda_U(B)$ for all B in $\mathscr{B}^\delta(U)$.

The result just established applies also to the member $T[B]$ of $\mathscr{B}^\delta(V)$ and the map $T^{-1}: V \to U$, so $\Lambda_U(T^{-1}[T[B]]) \leqq sup\{|\det(T^{-1})'(c)|: c \in T[B]\}\Lambda_V(T[B])$, $det(T^{-1})'(c) = (det\,T'(T^{-1}(c)))^{-1}$ for c in $T[B]$, and therefore $\Lambda_U(B) \leqq (1/inf\{|\det T'(b)|: b \in B\})\Lambda_V(T[B])$. Thus $\Lambda_U(B)\,inf_B|\det T'| \leqq \Lambda_V(T[B]) \leqq \Lambda_U(B)\,sup_B|\det T'|$. ■

18 THEOREM *If T is a continuously differentiable one to one map with non-vanishing Jacobian on an open subset U of $\mathbb{R}^p$ and $V = T[U] \subset \mathbb{R}^p$, then $\Lambda_V(T[B]) = \int_B |\det T'(x)|\,d\Lambda_U x$ for each member B of $\mathscr{B}^\delta(U)$.*

PROOF Suppose $B \in \mathscr{B}^\delta(U)$ and $B_{ki} = \{x \in B: (i-1)/k \leqq |det\,T'(x)| < i/k\}$ for positive integers k and i. Then $B_{ki} \in \mathscr{B}^\delta(U)$ for each k and i and in view of the preceding lemma, $\Lambda_V(T[B_{ki}])$ lies between $((i - 1/k)\Lambda_U(B)$ and $(i/k)\Lambda_U(B)$. It is evident that $\int_{B_{ki}} |det\,T'(x)|\,d\Lambda_U x$ also lies between the same two bounds for each k and i. Since $B = \bigcup_i B_{ki}$ for each k it follows that $|\Lambda_V(T[B]) - \int_B |det\,T'(x)|\,d\Lambda_U x| \leqq \sum_i (1/k)\Lambda_U(B_{ki}) = (1/k)\Lambda_U(B)$ and consequently $\Lambda_V(T[B]) = \int_B |det\,T'(x)|\,d\Lambda_U x$. ■

There are immediate consequences of the foregoing. The **Borel image** $\boldsymbol{T_{\mathscr{B}}\Lambda_U}$ is given by $T_{\mathscr{B}}\Lambda_U(B) = \Lambda_U(T^{-1}[B])$ for each B in $\mathscr{B}^\delta(V)$, and by the preceding theorem, $\Lambda_U(T^{-1}[B]) = \int_B |det(T^{-1})'| \, d\Lambda_V$. Consequently the Borel image $T_{\mathscr{B}}\Lambda_U$ is the indefinite integral $|det(T^{-1})'|.\Lambda_V$.

The "change of variable formula" is also a consequence. If $C \in \mathscr{B}^\delta(V)$ then $\int \chi_C \, d\Lambda_V = \Lambda_V(T[T^{-1}[C]]) = \int_{T^{-1}[C]} |det\, T'| \, d\Lambda_U = \int (\chi_C \circ T) \times |det\, T'| \, d\Lambda_U$, and it follows that $\int f \, d\Lambda_V = \int (f \circ T) |det\, T'| \, d\Lambda_U$ for all f in $L_1(\Lambda_V)$. Thus:

19 Corollary *If T is a continuously differentiable one to one map with non-vanishing Jacobian on an open subset U of $\mathbb{R}^p$ and $V = T[U]$, then*

(i) *the Borel image $T_{\mathscr{B}}\Lambda_U$ is the indefinite integral $|\det(T^{-1})'|.\Lambda_V$ and*
(ii) $\int f \, d\Lambda_V = \int (f \circ T) |\det T'| \, d\Lambda_U$ *for all f in $L_1(\Lambda_V)$.*

NOTES

(i) A form of the preceding results holds for maps $T: U \to V$ that are regular, in the sense that $T^{-1}[K]$ is compact for each compact subset K of V, but are not necessarily one to one. If T is regular and $|det\, T'|$ is non-vanishing, then T is locally one to one by the implicit functions theorem, and it can be shown that each f in $C_c(V)$ is the sum of finitely many members $f_1, \ldots f_n$ such that $f_i \circ T$ has a compact support K_i and T is one to one on an open set U_i containing K_i. Corollary 19 can then be used to conclude that $\int f_i \, d\Lambda_V = \int |det\, T'| f_i \circ T \, d\Lambda_U$ for each i, and hence the same equality for f.

Part (i) of the corollary takes a slightly different form in this case: if for y in V, $\varphi(y) = \sum \{1/|det\, T'(x)| : x \textit{ such that } T(x) = y\}$, then $T_{\mathscr{B}}\Lambda_U = \varphi.\Lambda_V$.

(ii) If $|det\, T'|$ vanishes at points of U, then $T_{\mathscr{B}}\Lambda_U$ may fail to be an indefinite integral. For example: if T is a constant c on a compact set A and $\Lambda_U(A) \neq 0$, then $T_{\mathscr{B}}\Lambda_U(\{c\}) \geqq \Lambda_U(A) \neq 0$, and $T_{\mathscr{B}}\Lambda_U$ cannot be an indefinite integral w.r.t. Λ_V.

(iii) The hypothesis "T is continuously differentiable" can be weakened substantially. See for example W. Rudin [1] or K. T. Smith [1].

*SUPPLEMENT: MAPS OF BOREL MEASURES**; CONVOLUTION*

Suppose X is a locally compact Hausdorff space, $\mu: \mathscr{B}^\delta(X) \to \mathbb{R}^+$ is a Borel measure for X and $\mu_\#$ is the canonical extension of μ to $\mathscr{L}\mathscr{B}^\delta(X)$. That is, $\mu_\#(B) = sup\{\mu(A): A \in \mathscr{B}^\delta(X)$ and $A \subset B\}$ for each member B of $\mathscr{L}\mathscr{B}^\delta(X)$. Then $\mu_\#$ is a measure* (a measure in the extended sense), and if μ is a regular measure then $\mu_\#$ is **inner regular** in the sense that $\mu_\#(B) = sup\{\mu_\#(A)$: A a compact subset of $B\}$ for each B in $\mathscr{L}\mathscr{B}^\delta(X)$. On the other hand, an inner regular measure* ν

on $\mathscr{L}\mathscr{B}^\delta(X)$ that is finite on compact sets is evidently the canonical extension of the regular Borel measure $\nu|\mathscr{B}^\delta(X)$.

Suppose that X and Y are locally compact Hausdorff spaces and that T is a continuous map of X into Y. If B is a compact subset of Y, then $T^{-1}[B]$ is closed and therefore locally $\mathscr{B}^\delta(X)$ measurable, and consequently every compact set belongs to the σ-field $\{A: T^{-1}[A]$ is in $\mathscr{L}\mathscr{B}^\delta(X)\}$. Hence $T^{-1}[B] \in \mathscr{L}\mathscr{B}^\delta(X)$ for each B in $\mathscr{B}^\delta(Y)$.

The **Borel image $T_{\mathscr{B}}\mu$**, or just **$T\mu$**, of an inner regular measure* μ on $\mathscr{B}^\delta(X)$ under a continuous map $T: X \to Y$ is defined by $T\mu(B) = sup\{\mu(A): A \in \mathscr{B}^\delta(X)$ and $A \subset T^{-1}[B]\}$ for each member B of $\mathscr{B}^\delta(Y)$. If μ is inner regular, then for B in $\mathscr{B}^\delta(Y)$ there is a compact subset D of $T^{-1}[B]$ such that $\mu(D)$ is near $T\mu(B)$, so $T[D]$ is a compact subset of B and $T\mu(T[D]) \geqq \mu(D)$, and hence $T\mu$ is inner regular. Thus the Borel image of an inner regular measure* is inner regular.

If μ is a regular Borel measure for X and $T: X \to Y$ is continuous, then the Borel image measure* $T\mu$ is given by $T\mu(B) = \mu_{\#}(T^{-1}[B])$ for each B in $\mathscr{B}^\delta(Y)$. In this case lemma 2 applies directly and yields the following.

20 PROPOSITION *Suppose X and Y are locally compact Hausdorff spaces, $T: X \to Y$ is continuous, μ is a regular Borel measure for X and $T\mu$ is its Borel image measure*.*

Then a $\mathscr{B}^\delta(Y)$ σ-simple function f belongs to $L_1(T\mu)$ iff $f \circ T$ is in $L_1(\mu_{\#})$, or iff $f \circ T$ agrees locally μ a.e. with a member of $L_1(\mu)$, and in this case $\int f\, dT\mu$ is $\int f \circ T\, d\mu_{\#}$, which is equal to $\int f \circ T\, d\mu$ iff $f \circ T$ has a σ-compact support.

There is no serious difficulty in extending the preceding proposition to the integral* (integral in the extended sense) w.r.t. $T\mu$, but one has to be a little careful about infinities even if X is σ-compact. Let **$L^+(Y)$** be the class of countable linear combinations with non-negative coefficients of characteristic functions of members of $\mathscr{B}^\delta(Y)$. These functions can also be described as the non-negative locally $\mathscr{B}^\delta(Y)$ measurable $\mathbb{R}^*$ valued functions with σ-compact support, and each of them is integrable* w.r.t. every Borel measure for Y. But this is not necessarily the case for a measure* $T\mu$. (E.g., let $X = \mathbb{Z} \times \mathbb{Z}$, $Y = \mathbb{Z}$, $T(p,q) = p + q$ for (p,q) in X, and let μ be counting measure. Then $T\mu(A) = \infty$ for every non-empty subset A of Y, and only the zero function is $T\mu$ integrable*.)

21 PROPOSITION *Suppose μ is a regular Borel measure for X, $T: X \to Y$ is continuous and X is σ-compact.*

If $f \in L^+(Y)$, then $f \circ T$ is μ integrable, and if $\int f \circ T\, d\mu < \infty$, then f is also $T\mu$ integrable*. If f is $T\mu$ integrable*, then $f \circ T$ is μ integrable* and $\int f\, dT\mu = \int f \circ T\, d\mu$.*

PROOF The function $f \circ T$ is non-negative, locally $\mathscr{B}^\delta(X)$ measurable and has a σ-compact support, so $f \circ T = \sum_n a_n \chi_{A_n}$ with $a_n \geqq 0$ and $A_n \in \mathscr{B}^\delta(X)$ for each n. Hence $f \circ T$ is μ integrable*.

If $\int f \circ T\, d\mu < \infty$ and B is a member of $\mathscr{B}^\delta(Y)$ such that $m = inf_{y \in B} f(y) > 0$, then $\infty > \int f \circ T\, d\mu \geqq \int (m\chi_B) \circ T\, d\mu = mT\mu(B)$. But f is a countable linear combination with non-negative coefficients of characteristic functions of such sets B of finite $T\mu$ measure, so f is $T\mu$ integrable*.

If f is $T\mu$ integrable*, then $f = \sum_n b_n \chi_{B_n}$ with $b_n \geqq 0$, $B_n \in \mathscr{B}^\delta(Y)$ and $T\mu(B_n) < \infty$ for all n. Then $b_n \chi_{B_n}$ is $T\mu$ integrable and $\int b_n \chi_{B_n}\, dT\mu = \int b_n \chi_{B_n} \circ T\, d\mu$ by the preceding proposition. Summing on n then shows that $\int f\, dT\mu = \int f \circ T\, d\mu$. ■

Let us suppose that G is a locally compact Hausdorff topological group, and that Γ: $G \times G \to G$ is the group map given by $\Gamma(x, y) = xy$ for all x and y in G. The **convolution** $\mu \star \nu$ of regular Borel measures μ and ν for G is defined to be the Borel image measure* $\Gamma\mu \otimes_{\mathscr{B}} \nu$. Thus for A in $\mathscr{B}^\delta(G)$, $\mu \star \nu(A) = (\mu \otimes_{\mathscr{B}} \nu)_{\#}(\{x, y): xy \in A\}) = sup\{\mu \otimes_{\mathscr{B}} \nu(B)$: $B \in \mathscr{B}^\delta(G \times G)$ and $\Gamma[B] \subset A\}$. We can compute $\mu \star \nu(A)$ as an iterated integral $\int\int \chi_A(xy)\, d\mu x\, d\nu y$. In case G is σ-compact, $f \circ \Gamma$ is $\mu \otimes_{\mathscr{B}} \nu$ integrable* for f in $L^+(G)$ according to the preceding proposition. The Tonelli theorem 7.14 then applies and it exhibits $\int f \circ \Gamma\, d\mu \otimes_{\mathscr{B}} \nu$ as an iterated integral.

Explicitly: for each x and y, $x \mapsto f(x, y)$ is μ integrable*, $y \mapsto f(x, y)$ is ν integrable*, $y \mapsto \int f(x, y)\, d\mu x$ is ν integrable*, $x \mapsto \int f(x, y)\, d\nu y$ is μ integrable*, and the integral in the extended sense of f w.r.t. $\mu \otimes_{\mathscr{B}} \nu$ is identical with the two iterated integrals. This fact, together with the preceding proposition, establishes the following.

22 PROPOSITION *If μ and ν are regular Borel measures for a σ-compact G and $f \in L^+(G)$, then $\int f \circ \Gamma\, d\mu \otimes_{\mathscr{B}} \nu = \int\int f(xy)\, d\mu x\, d\nu y = \int\int f(xy)\, d\nu y\, d\mu x$.*

If f is $\mu \star \nu$ integrable, then $\int f\, d\mu \star \nu = \int f \circ \Gamma\, d\mu \otimes_{\mathscr{B}} \nu$. If $\int f \circ \Gamma\, d\mu \otimes_{\mathscr{B}} \nu < \infty$, then f is $\mu \star \nu$ integrable*.*

Here is an example of the use of these results. Suppose μ, ν and η are regular Borel measures for the σ-compact group G, η is left invariant, $f \in L^+(G)$ and h is a member of $L^+(G)$ such that $\nu = h.\eta$ (that is, $\nu(A) = \int_A h\, d\eta$ for each A in $\mathscr{B}^\delta(G)$—whence $\int g\, d\nu = \int gh\, d\eta$ for all g in $L^+(G)$). Then $\int f \circ \Gamma\, d\mu \otimes_{\mathscr{B}} \nu = \int\int f(xy) h(y)\, d\eta y\, d\mu x$, and since η is left invariant, this is $\int\int f(y) h(x^{-1} y)\, d\eta y\, d\mu x$. Consequently the function $y \mapsto \int h(x^{-1} y)\, d\mu x$, which we denote $\boldsymbol{\mu \star h}$, belongs to $L^+(G)$ and $\int (\mu \star h) f\, d\eta = \int f \circ \Gamma\, d\mu \otimes_{\mathscr{B}} \nu$. If f is $\mu \star \nu$ integrable*, and in particular if $\int f \circ \Gamma\, d\mu \otimes_{\mathscr{B}} \nu < \infty$, then $\int (\mu \star h) f\, d\eta = \int f\, d\mu \star \nu = \int f\, d\mu \star (h.\eta)$.

If, further, $\mu = g.\eta$ for some g in $L^+(G)$, then $\int f \circ \Gamma \, d\mu \otimes_{\mathscr{B}} \nu = \int\int f(y)g(x)h(x^{-1}y)\, d\eta x \, d\eta y$ and so the function $y \mapsto \int g(x)h(x^{-1}y)\, d\eta x$, denoted $\boldsymbol{g \star_\eta h}$, is in $L^+(G)$ and $\int (g \star_\eta h) f \, d\eta = \int f \circ \Gamma \, d\mu \otimes_{\mathscr{B}} \nu$. If f is $\mu \star \nu$ integrable*, and in particular if $\int f \circ \Gamma \, d\mu \otimes_{\mathscr{B}} \nu < \infty$, then $\int f \, d\mu \star \nu = \int f \, d(g.\eta) \star (h.\eta) = \int (g \star_\eta h) f \, d\eta$. If $\mu \star \nu(A) < \infty$ for all A in $\mathscr{B}^\delta(G)$, then $\mu \star (h.\eta) = (\mu \star h).\eta$ and $(g.\eta) \star (h.\eta) = (g \star_\eta h).\eta$.

We record these facts, prefacing the statement with a simple result about indefinite integrals w.r.t. a regular Borel measure μ for a locally compact Hausdorff space X.

Let $\|\mu\|_V = \mu_{\#}(X) = sup\{\mu(A) \colon A \in \mathscr{B}^\delta(X)\}$. The measure μ is **totally finite** iff $\|\mu\|_V < \infty$.

23 Proposition *Suppose μ is a regular Borel measure for a locally compact Hausdorff space X and $f \in L^+(X)$. Then $f.\mu$ is a regular Borel measure if f is locally μ integrable, and if f is μ integrable, then $f.\mu$ is totally finite and $\|f.\mu\|_V = \|f\|_1$.*

PROOF We show only that the indefinite integral of a locally μ integrable function f is regular. For $e > 0$ and A in $\mathscr{B}^\delta(X)$ choose n so $\int_A (f - f \wedge n)\, d\mu < e$ and choose a compact subset B of A so that $\mu(A \setminus B) < e/n$. Then $\int_A f \, d\mu - \int_B f \, d\mu = \int_{A\setminus B}(f - f \wedge n)\, d\mu + \int_{A\setminus B} f \wedge n \, d\mu \leqq \int_A (f - f \wedge n)\, d\mu + \int_{A\setminus B} n \, d\mu < e + n(e/n)$, so $f.\mu$ is regular. ■

Here is a summary of some immediate consequences of the foregoing results and the definition of $\star$. We leave details of the proof to the readers.

24 Theorem (Summary) *Suppose G is a σ-compact group; f, g and h belong to $L^+(G)$; μ, ν and η are regular Borel measures for G; η is left invariant; and g and h are locally η integrable*.*

(i) *If $\|\mu \star \nu\|_V$ is finite, then it is a regular Borel measure, and in any case, $\|\mu \star \nu\|_V = \|\mu\|_V \|\nu\|_V$.*
(ii) *If f is $\mu \star (h.\eta)$ integrable*, so if $\int\int f(xy)h(y)\, d\mu x \, d\eta y < \infty$, then $\int f(\mu \star h)\, d\eta = \int f \, d\mu \star h.\eta$.*
(iii) *If f is $(g.\eta) \star (h.\eta)$ integrable*, so if $\int\int f(xy)\, g(x) h(y)\, d\eta x \, d\eta y$ is finite, then $\int f \, d(g.\eta) \star (h.\eta) = \int f(g \star_\eta h)\, d\eta$.*
(iv) *If h belongs to $L_1(\eta)$ and $\|\mu\|_V < \infty$, then $\mu \star h \in L_1(\eta)$, $\|\mu \star h\|_1 = \|\mu\|_V \|h\|_1$, and $\mu \star (h.\eta) = (\mu \star h).\eta$. If g also belongs to $L_1(\eta)$, then $(g \star_\eta h) \in L_1(\eta)$ and $(g \star_\eta h).\eta = (g.\eta) \star (h.\eta)$.*

Chapter 9

SIGNED MEASURES AND INDEFINITE INTEGRALS

A real (finite) valued function f on X is locally μ integrable iff μ is a measure on a δ-ring $\mathscr{A}$ of subsets of X and $f\chi_A \in L_1(\mu)$ for all A in $\mathscr{A}$. In this case $f.\mu$, the indefinite integral of f with respect to μ, is the function $A \mapsto \int_A f\,d\mu$ for A in $\mathscr{A}$. This function is always countably additive and hence $f.\mu$ is a measure if f is non-negative. Consequently $f.\mu$ is the difference of two measures, $f^+.\mu$ and $f^-.\mu$. These two measures have little to do with each other: one of them "lives on" the set $\{x: f(x) \geqq 0\}$ and the other lives on $\{x: f(x) < 0\}$.

There are several natural questions. Which countably additive real valued functions on $\mathscr{A}$ are differences of measures? If a function is the difference of measures, may these measures be chosen so that they live on disjoint sets? Which countably additive real valued functions are indefinite integrals? This section is devoted to an investigation of these and related questions.

A **signed measure** is a real (finite) valued, countably additive function on a δ-ring. Suppose μ is a signed measure on a δ-ring $\mathscr{A}$ of subsets of X. A subset B of X is ***μ* positive** iff B is locally $\mathscr{A}$ measurable (i.e., $A \cap B \in \mathscr{A}$ for all A in $\mathscr{A}$) and $\mu(A) \geqq 0$ for all A in $\mathscr{A}$ that are subsets of B. There are alternative descriptions: for each A in $\mathscr{A}$, $A \cap B \in \mathscr{A}$ and $\mu(A \cap B) \geqq 0$; or B is locally $\mathscr{A}$ measurable and $A \mapsto \mu(A \cap B)$ is a measure. Evidently each member A of $\mathscr{A}$ that is a subset of a μ positive set is μ positive, and the union of countably many μ positive sets is μ positive.

A set B is ***μ* negative** iff it is locally $\mathscr{A}$ measurable and $\mu(A) \leqq 0$ for each member A of $\mathscr{A}$ that is a subset of B, or iff B is $-\mu$ positive.

1 LEMMA *If μ is a signed measure on $\mathscr{A}$, $A \in \mathscr{A}$ and $\mu(A) > 0$ then A contains a μ positive set B with $\mu(B) \geqq \mu(A)$.*

PROOF We assert that for each $e > 0$ there is a member C_0 of $\mathscr{A}$ such that $C_0 \subset A$, $\mu(C_0) \leqq 0$ and each member D of $\mathscr{A}$ with $D \subset A \setminus C_0$ has μ measure greater than $-e$. If this were not the case, then one could select recursively a disjoint sequence $\{D_n\}_n$ in $\mathscr{A}$ with $D_n \subset A$ and $\mu(D_n) \leqq -e$ for all n. But every δ-ring is closed under dominated countable union and $\bigcup_n D_n \subset A$, So $\bigcup_n D_n \in \mathscr{A}$ and $\mu(\bigcup_n D_n) = \sum_n \mu(D_n) = -\infty$, and this is a contradiction.

Select recursively, C_{n+1} in $\mathscr{A}$ so that $C_{n+1} \subset A \setminus \bigcup_{k=0}^{n} C_k$, $\mu(C_{n+1}) \leqq 0$ and each member D of $\mathscr{A}$ with $D \subset A \setminus \bigcup_{k=0}^{n+1} C_k$ has μ measure $> -2^{-n-1}$. The argument of the preceding paragraph shows that this selection is always possible. Then $B = A \setminus \bigcup_n C_n$ is μ positive, and $\mu(B) = \mu(A) - \sum_n \mu(C_n) \geqq \mu(A)$. ■

We recall that if $\mathscr{A}$ is a family of sets, then $\mathscr{A}_\sigma$ is the family of all unions of countably many members of $\mathscr{A}$. The **upper variation μ^+** of a signed measure μ on $\mathscr{A}$ is defined, for A in $\mathscr{A}$, by $\mu^+(A) = sup\{\mu(B)$: $B \subset A$ and $B \in \mathscr{A}\}$ and the **lower variation μ^-** is given by $\mu^-(A) = -inf\{\mu(B)$: $B \subset A$ and $B \in \mathscr{A}\} = (-\mu)^+$. If a member A of $\mathscr{A}$ is μ positive, then $\mu^+(A) = \mu(A)$ and if A is μ negative then $\mu^-(A) = -\mu(A)$. If a set A is the union of a μ positive set A^+ and a μ negative set A^- and $A^+ \cap A^- = \varnothing$, then $\{A^+, A^-\}$ is a **Hahn decomposition** of A relative to μ. A representation of μ as the difference $\nu - \rho$ of measures ν and ρ is a **Jordan representation.**

2 THEOREM (JORDAN, HAHN) *If μ is a signed measure on $\mathscr{A}$, then the upper and lower variations of μ are measures and μ is their difference. Moreover, each member A of $\mathscr{A}_\sigma$ is the union of a μ positive set A^+ and a μ negative set A^- that is disjoint from it.*

PROOF For A in $\mathscr{A}$, choose a sequence $\{B_n\}_n$ of members of $\mathscr{A}$ that are subsets of A such that $\mu^+(A) = sup_n \mu(B_n)$, and for each n choose, using lemma 1, a μ positive subset C_n of B_n with $\mu(C_n) \geqq \mu(B_n)$. Let $A^+ = \bigcup_n C_n$. Then $A^+ \in \mathscr{A}$ and since $C_n \subset A^+ \subset A$ for each n, $\mu^+(A) \leqq \mu^+(A^+) \leqq \mu^+(A)$, so $\mu^+(A) = \mu^+(A^+)$. But A^+ is the union $\bigcup_n C_n$ of μ positive sets and is therefore μ positive, so $\mu^+(A^+) = \mu(A^+)$. Thus $\mu(A^+) = \mu^+(A)$.

Let $A^- = A \setminus A^+$. Clearly A^- is μ negative and $\mu(A^-) = \mu(A) - \mu^+(A) = inf\{\mu(B)$: $B \subset A$ and $B \in \mathscr{A}\} = -\mu^-(A)$.

If $\{A_n\}_n$ is a disjoint sequence in $\mathscr{A}$, then $\bigcup_n A_n^+$ and $\bigcup_n A_n^-$ furnish a Hahn decomposition for $\bigcup_n A_n$, and the last assertion of the theorem follows. If $\bigcup_n A_n \in \mathscr{A}$, then $\sum_n \mu^+(A_n) = \sum_n \mu(A_n{}^+) = \mu(\bigcup_n A_n{}^+) = \mu^+(\bigcup_n A_n)$ so μ^+ is a measure, and so is $\mu^- = (-\mu)^+$. Hence $\mu = \mu^+ - \mu^-$ is a Jordan representation of μ. ■

It is worth noticing that if μ is a signed measure on $\mathscr{A}$ and the underlying space X belongs to $\mathscr{A}_\sigma$ (as is the case for Borel measures for $\mathbb{R}^n$), then X has a Hahn decomposition into a μ positive and a μ negative set.

The class $\boldsymbol{M(\mathscr{A})}$ of all signed measures on a δ-ring $\mathscr{A}$ is a vector space, and it is partially ordered by agreeing that, for μ and λ in $M(\mathscr{A})$, $\lambda \geqq \mu$ iff $\lambda(A) \geqq \mu(A)$ for all A in $\mathscr{A}$. Suppose that $\lambda \geqq \mu$, $A \in \mathscr{A}$ and that $\{A^+, A^-\}$ is a Hahn decomposition of A for μ. If $\lambda \geqq \mu$ and $\lambda \geqq 0$, then $\lambda(A) = \lambda^+(A) = \lambda(A^+) \geqq \mu(A^+) = \mu^+(A)$. Thus $\lambda \geqq \mu^+$, and we have shown that μ^+ is the smallest signed measure that is greater than or equal both 0 and μ. Thus, μ^+ is the supremum of $\{0, \mu\}$ relative to the ordering of $M(\mathscr{A})$. Consequently $M(\mathscr{A})$, with this ordering, is a vector lattice, and that $\mu \vee \nu = \nu + (\mu - \nu) \vee 0 = \nu + (\mu - \nu)^+$. In particular, the definition of μ^+ is consistent with the usual lattice convention: $\mu^+ = \mu \vee 0$.

3 CAUTION Suppose μ and ν are measures on $\mathscr{A}$. It is *not* always the case that $\mu \vee \nu(A) = max\{\mu(A), \nu(A)\} = \mu(A) \vee \nu(A)$. For example: if $X = \{0, 1\}$, $\mathscr{A} = \{\varnothing, \{0\}, \{1\}, \{0, 1\}\}$, μ is unit mass at 0 and ν is unit mass at 1, then $\mu\{0\} \vee \nu\{0\} = \mu\{1\} \vee \nu\{1\} = \mu\{0, 1\} \vee \nu\{0, 1\} = 1$, but the smallest measure greater than or equal to each of μ and ν is $\mu + \nu$, which is 2 at $\{0, 1\}$. What is true is that $\mu \vee \nu(A) \geqq \mu(A) \vee \nu(A)$ for all A in $\mathscr{A}$—indeed, $\mu \vee \nu$ is the smallest measure for which this is the case. ■

A Jordan representation of a signed measure λ on $\mathscr{A}$ as the difference $\mu - \nu$ of measures μ and ν is not unique—one may add any measure on $\mathscr{A}$ to both μ and ν and get another Jordan representation. The **canonical Jordan representation** as the difference $\lambda^+ - \lambda^-$ of the variations has a special property: λ^+ and λ^- live, at least locally, on disjoint sets. Let us make this assertion precise.

Suppose μ is a measure on $\mathscr{A}$. We agree that **μ lives on a set C, (μ is carried by C, C is a carrier for μ)** iff $A \cap C \in \mathscr{A}$ and $\mu(A) = \mu(A \cap C)$ for all A in $\mathscr{A}$. Examples: If the underlying space X has a Hahn decomposition $X = X^+ \cup X^-$ for a signed measure μ on $\mathscr{A}$, then μ^+ lives on X^+ and μ^- lives on X^-; if f is a locally $\mathscr{A}$ measurable non-negative function such that $f\chi_A \in L_1(\mu)$ for each A in $\mathscr{A}$ and if S is a locally $\mathscr{A}$ measurable support for f, then the indefinite integral $f.\mu$ lives on S. Evidently a measure μ lives on C iff $X \setminus C$ is locally of μ

measure zero in the sense that $A \cap (X \setminus C) \in \mathscr{A}$ and $\mu(A \cap (X \setminus C)) = 0$ for all A in $\mathscr{A}$.

A measure ν on $\mathscr{A}$ is said to be **singular w.r.t. a measure μ** on $\mathscr{A}$ iff ν lives on a set C that is locally of μ measure zero. This is the case iff μ is singular w.r.t. ν because μ lives on the set $X \setminus C$, which is locally of ν measure 0, and consequently μ and ν have disjoint carriers. We agree that μ and ν are **mutually singular**, and write $\mu \perp \nu$, iff μ and ν have disjoint carriers. The measures μ and ν are **locally** $\perp$, and μ is **locally** $\perp$ to ν, iff for each A in $\mathscr{A}$ there are disjoint members A_μ and A_ν of $\mathscr{A}$ whose union is A such that $\nu(A_\mu) = \mu(A_\nu) = 0$. The measure ν on $\mathscr{A}$ is **absolutely continuous** w.r.t. the measure μ on $\mathscr{A}$, $\nu \prec \mu$, iff $\nu(A) = 0$ whenever $\mu(A) = 0$ (we extend the definition on p. 113.)

4 LEBESGUE DECOMPOSITION THEOREM *Suppose μ and ν are measures on a δ-ring $\mathscr{A}$ of subsets of X, that $\nu_s(A) = \sup\{\nu(B): B \subset A, B \in \mathscr{A}$ and $\mu(B) = 0\}$ for all A in $\mathscr{A}$ and that $\nu_c = \nu - \nu_s$. Then ν_s and ν_c are measures, ν_s is locally $\perp$ to ν_c and to μ, and $\nu_c \prec \mu$.*

If $X \in \mathscr{A}_\sigma$, then $\nu_s \perp \mu$.

PROOF For each A in $\mathscr{A}$, the family $\{B: B \subset A, B \in \mathscr{A}$ *and* $\mu(B) = 0\}$ is closed under countable union and hence has a member A_s with $\nu(A_s) = sup\{\nu(B): B \in \mathscr{A}, B \subset A$ and $\mu(B) = 0\} = \nu_s(A)$. It follows that each subset of $A \setminus A_s$ of μ measure zero must have ν measure zero, whence $\nu_s(A \setminus A_s) = 0$, and ν_s and μ are locally $\perp$.

If $\mu(A) = 0$ then A_s may be taken to be A, so $\nu_c(A) = \nu(A) - \nu_s(A) = \nu(A) - \nu(A_s) = 0$. ∎

The measure ν_s is called the **singular part** of ν (w.r.t. μ) and ν_c is the **absolutely continuous** part. Of course there is also a Lebesgue decomposition of μ into singular part μ_s of μ (w.r.t. ν) and absolutely continuous part μ_c. But the immediate consequences of this fact—as well as theorem 4 itself—are corollaries of the following result.

5 FUNDAMENTAL LEMMA *If μ and ν are measures on a δ-ring $\mathscr{A}$ of subsets of X and $E \in \mathscr{A}$, then there is an $\mathscr{A}$ σ-simple function h, $0 \leqq h \leqq \chi_E$ such that $\mu(A) = h.(\mu + \nu)(A)$ and $\nu(A) = (1 - h).(\mu + \nu)(A)$ for all subsets A of E that belong to $\mathscr{A}$.*

If $E_0 = \{x: h(x) = 0\}$, $E_1 = \{x: h(x) = 1\}$, $E_c = E \setminus (E_0 \cup E_1)$ and $A \subset E_c$, then $\mu(E_0) = 0 = \nu(E_1)$, $\nu(A) = \int_A ((1 - h)/h)\, d\mu$ and $\mu(A) = \int_A ((h/(1 - h)\, d\nu$.

PROOF For the measure $\mu + \nu$ the class $L_2(\mu + \nu)$ of $\mathscr{A}$ σ-simple real valued functions f such that $|f|^2 \in L_1(\mu + \nu)$, together with the inner product given by $\langle f, g \rangle = \int fg\, d(\mu + \nu)$, is a Hilbert space (see chapter

6). If $f \in L_2(\mu + \nu)$ and $E \in \mathscr{A}$ then $|f\chi_E|^2$ is $(\mu + \nu)$ integrable and vanishes off a set of finite measure, so $|f\chi_E|$ is $(\mu + \nu)$ and hence μ integrable. Moreover $\int |f\chi_E|\,d\mu \leqq \int |f\chi_E|\,d(\mu + \nu) \leq \|f\|_2 \|\chi_E\|_2$ by the Cauchy–Schwartz inequality and consequently $F = (f \mapsto \int_E f\,d\mu)$ is a bounded linear functional on $L_2(\mu + \nu)$. According to the Riesz theorem 6.12, there is h in $L_2(\mu + \nu)$ such that $F(f) = \int fh\,d(\mu + \nu)$ for all f in $L_2(\mu + \nu)$. Since $F(f) = F(f\chi_E) = \int_E f\chi_E h\,d(\mu + \nu)$, we may suppose $h = 0$ outside E, and since $F(f) \geqq 0$ for $f \geqq 0$, $h \geqq 0$ $(\mu + \nu)$ a.e., and we may suppose $h \geqq 0$. If $e > 0$ and $B = \{x\colon x \in E \textit{ and } h(x) > 1 + e\}$, then $\mu(B) = F(\chi_B) = \int_E \chi_B h\,d(\mu + \nu) \geqq (1 + e)(\mu + \nu)(B)$, whence $(\mu + \nu)(B) = 0$ and we may suppose that $0 \leqq h \leqq \chi_E$. Since $\mu(A) = h.(\mu + \nu)(A)$, $\nu(A) = \int_A 1\,d(\mu + \nu) - \mu(A) = \int_A (1 - h)\,d(\mu + \nu)$ for each A in $\mathscr{A}$ that is a subset of E.

If $f \in L_2(\mu + \nu)$ and $A \subset E_c$, then $\int_A f\,d\mu = \int_A fh\,d(\mu + \nu)$ and so $\int_A f(1 - h)\,d\mu = \int_A fh\,d\nu$. If $A_n = \{x\colon 1/n < h \leqq 1 - 1/n\}$, then both $\chi_{A_n}/(1 - h)$ and χ_{A_n}/h belong to $L_2(\mu + \nu)$, and we infer that $\mu(A_n) = \int \chi_{A_n}(h/1 - h)\,d\nu$ and $\nu(A_n) = \int \chi_{A_n}(1 - h)/h\,d\mu$. Consequently, $\mu(A) = \lim_n \mu(A_n) = \int_A (h/(1 - h))\,d\nu$ and $\nu(A) = \lim_n \nu(A_n) = \int_A (1 - h)/h\,d\mu$.

■

6 Theorem (Radon–Nikodym) *If μ and ν are measures on a δ-ring $\mathscr{A}$ of subsets of X, $X \in \mathscr{A}_\sigma$ and $\nu \prec \mu$, then $\nu = r.\mu$ for some non-negative, $\mathscr{A}$ σ-simple function r on X.*

PROOF Suppose $X = \bigcup_n E_n$ and $\{E_n\}_n$ is a disjoint sequence in $\mathscr{A}$. Then according to the preceding lemma, for each n there is a function h_n, $0 \leqq h_n \leqq \chi_{E_n}$ such that $\mu(A \cap E_n) = \int_{A \cap E_n} h_n\,d(\mu + \nu)$ iff $A \in \mathscr{A}$. The subset Z of $A \cap E_n$ on which h_n vanishes has μ measure 0 and hence ν measure zero, and so $\mu(A \cap E_n) = \int_{A \cap E_n} (h_n + \chi_Z)\,d(\mu + \nu)$. In brief, we may suppose h_n is strictly positive. In this case, according to the preceding lemma, if $r_n = (1 - h_n)/h_n$, then $\nu(A \cap E_n) = r_n.\mu(A)$. If $r = \sum_n r_n$, then $\nu = r.\mu$. ■

NOTE An alternative proof, independent of the Riesz representation theorem 6.12, can be based directly on lemma 1. But the argument given here, which is due to von Neumann, is prettier. ■

There is an extension of the preceding theorem. Suppose $\mathscr{A}$ is a δ-ring of subsets of X and f is a real valued locally measurable function on X, *not necessarily locally μ integrable.* **The indefinite integral $f.\mu$** of f w.r.t. μ is defined on the class $\mathscr{B}$ of all members A of $\mathscr{A}$ such that $f\chi_A \in L_1(\mu)$ by setting $f.\mu(A) = \int_A f\,d\mu$.

7 Proposition *If μ is a measure on a δ-ring $\mathscr{A}$ of subsets of X and f is a locally $\mathscr{A}$ measurable real valued function on X, then the domain*

$\mathscr{B}$ of $f.\mu$ is a δ-ring, $\mathscr{B} \subset \mathscr{A} \subset \mathscr{B}_\sigma$, and $f.\mu$ is a signed measure on $\mathscr{B}$ that vanishes on all members of $\mathscr{B}$ with μ measure zero.

PROOF We show only that $\mathscr{A} \subset \mathscr{B}_\sigma$, and leave to the reader the rest of the proof. Suppose $A \in \mathscr{A}$ and that $B_n = \{x: x \in A, n < f(x) \leqq n + 1\}$ for each integer n. Then $B_n \in \mathscr{A}$ because f is locally $\mathscr{A}$ measurable, $|f\chi_{B_n}| \leqq |n| + 1$ so $f\chi_{B_n} \in L_1(\mu)$, and hence $B_n \in \mathscr{B}$. Consequently $A = \bigcup \{B_n: n \in \mathbb{Z}\} \in \mathscr{B}_\sigma$. ■

It is worth noticing a few of the simple consequences of the condition that $\mathscr{A}$ and $\mathscr{B}$ be δ-rings and $\mathscr{B} \subset \mathscr{A} \subset \mathscr{B}_\sigma$. First: if μ is a measure on $\mathscr{A}$ and $\nu = \mu|\mathscr{B}$ is its restriction to $\mathscr{B}$, then the integrals I_μ and I_ν are identical. Second: $\mathscr{B}$ is an **ideal** in $\mathscr{A}$ in the sense that $A \cap B \in \mathscr{B}$ if $A \in \mathscr{A}$ and $B \in \mathscr{B}$. This is the case because $A \cap B$ is the union of a countable number of members of $\mathscr{B}$ and the δ-ring $\mathscr{B}$ is closed under countable dominated union. It follows that if $\mathscr{C}$ is also a δ-ring and $\mathscr{C} \subset \mathscr{A} \subset \mathscr{C}_\sigma$, then $\{B \cap C: B \in \mathscr{B} \text{ and } C \in \mathscr{C}\}$ is identical with $\mathscr{B} \cap \mathscr{C}$. Lastly: if μ is a measure on a δ-ring $\mathscr{B}$ and ν is a measure on $\mathscr{C}$, then the integrals I_μ and I_ν are identical iff $\mu|\mathscr{B} \cap \mathscr{C} = \nu|\mathscr{B} \cap \mathscr{C}$ and $\mathscr{B} \cup \mathscr{C} \subset (\mathscr{B} \cap \mathscr{C})_\sigma$.

A signed measure ν on a δ-ring $\mathscr{B}$ is **absolutely continuous** w.r.t. a measure μ on $\mathscr{A}$, and we write $\boldsymbol{\nu \prec \mu}$, iff $\mathscr{B} \subset \mathscr{A} \subset \mathscr{B}_\sigma$ and $\nu(B) = 0$ for all members B of $\mathscr{B}$ with μ measure zero (or equivalently $\nu^+(B) = \nu^-(B) = 0$ for all such B). A signed measure ν is **an indefinite integral** w.r.t. μ iff there is a locally $\mathscr{A}$ measurable real valued function f such that $\nu = (f.\mu)|\mathscr{B}$. *The* indefinite integral $\nu = f.\mu$ of a locally μ integrable function f is *an* indefinite integral w.r.t. μ (no surprise) and $\nu \prec \mu$.

8 RADON–NIKODYM THEOREM FOR SIGNED MEASURES *If μ is a measure on a δ-ring $\mathscr{A}$ of subsets of X and $X \in \mathscr{A}_\sigma$, then every signed measure ν that is absolutely continuous w.r.t. μ is an indefinite integral w.r.t. μ.*

PROOF We may assume without loss of generality that ν is a measure because it is the difference $\nu^+ - \nu^-$ of measures on the domain $\mathscr{B}$ of ν that are absolutely continuous w.r.t. μ. Evidently ν is absolutely continuous w.r.t. $\mu|\mathscr{B}$ and since $\mathscr{A} \subset \mathscr{B}_\sigma$, $X \in \mathscr{A}_\sigma \subset \mathscr{B}_\sigma$. Hence, by theorem 6, $\nu = r.(\mu|\mathscr{B})$ for some $\mathscr{B}$ σ-simple non-negative real valued function r. The theorem follows. ■

We sketch without proof a few straightforward propositions about indefinite integrals. Suppose that μ is a measure on a δ-ring $\mathscr{A}$ of subsets of X and that r and s are locally μ integrable functions on X. Then $(r + s).\mu = r.\mu + s.\mu$. If $s \geqq 0$ so that $s.\mu$ is a measure, then $r.(s.\mu) = (rs).\mu$ and if rs is μ integrable, then $\int r\, d(s.\mu) = \int rs\, d\mu$. If s is never zero and $\mathscr{B}$ is the domain of $(1/s).\mu$, then $\mathscr{B} \subset \mathscr{A} \subset \mathscr{B}_\sigma$ and $s.(1/s).\mu = \mu | \mathscr{B}$.

If μ and ν are measures on $\mathscr{A}$ and r is a locally $\mathscr{A}$ measurable real valued function on the underlying space X such that $\nu = r.\mu$ then r is called a **Radon–Nikodym derivative** of ν w.r.t. μ. It is not unique, but two such derivatives differ by a function with a support of locally μ measure zero. Radon–Nikodym derivatives of ν w.r.t. μ are denoted (ambiguously) $\boldsymbol{d\nu/d\mu}$. If f is both μ and ν integrable, then $\int (f)(d\nu/d\mu)\,d\mu = \int f\,d\nu$. If μ also has a derivative with respect to ν, then $(d\mu/d\nu)(d\nu/d\mu) = 1$ locally μ a.e., and if ν has a derivative with respect to another measure π, then $(d\mu/d\nu)(d\nu/d\pi) = d\mu/d\pi$.

There is a natural definition of the integral of a function f w.r.t. a signed measure ν on $\mathscr{A}$. A function f is ν **integrable** iff $f = \sum_n a_n \chi_{A_n}$ for sequences $\{a_n\}_n$ in $\mathbb{R}$ and $\{A_n\}_n$ in $\mathscr{A}$ with $\sum_n |a_n| |\nu^+(A_n) + \nu^-(A_n)| < \infty$, and in this case $\int f\,d\nu = \sum_n a_n \nu(A_n) = \int f\,d\nu^+ - \int f\,d\nu^-$. Evidently $L_1(\nu) = L_1(\nu^+) \cap L_1(\nu^-)$. The measure $V_\nu = \nu^+ + \nu^-$ is called the **variation of $\boldsymbol{\nu}$**. It is easy to see that for A in $\mathscr{A}$, $V_\nu(A) = sup\{\nu(B) - \nu(C)$: $B, C \in \mathscr{A}$ and $B \subset A$, $C \subset A\} = sup\{\sum_{i=1}^k |\nu(B_i)|$: $\{B_i\}_{i=1}^k$ is a disjoint family of members of $\mathscr{A}$ that are subsets of $A\}$, and if $\nu = f.\mu$ for a measure μ then $V_\nu = |f|.\mu$.

Evidently $\nu \prec V_\nu$, *and if* $\nu = r.V_\nu$, then $\{x: r(x) > 0\}$ is a ν positive set, $\{x: r(x) \leqq 0\}$ is ν negative and consequently there is a Hahn decomposition of X for ν.

Thus the Radon–Nikodym theorem for V_ν implies the Hahn decomposition theorem for ν. There is a converse: if μ and ν are measures on $\mathscr{A}$, $\nu \prec \mu$ and for each real number k, there is a Hahn decomposition of X for $\mu - k\nu$, then ν is an indefinite integral w.r.t. μ.

NOTE The first result of this sort, that the Radon–Nikodym theorem for μ is equivalent to the proposition that the dual of $L_1(\mu)$ is $L_\infty(\mu)$, is due to I. Segal; related results were established by J. M. G. Fell and by one of the authors. See J. L. Kelley, *Math. Annalen* **163** (1966) 89–94, for details and further results.

SUPPLEMENT: DECOMPOSABLE MEASURES

The Radon–Nikodym theorem and the Hahn decomposition theorem were established in the foregoing chapter under the assumption that μ is a measure on a δ-ring $\mathscr{A}$ of subsets of X and $X \in \mathscr{A}_\sigma$. If X is a locally compact Hausdorff space and μ is a regular measure on the Borel δ-ring $\mathscr{B}^\delta(X)$, then $X \notin (\mathscr{B}^\delta(X))_\sigma$ unless X is σ-compact. Nevertheless the foregoing theorems still hold for μ. The principal fact needed for the proof is that there is a suitable disjoint subfamily of $\mathscr{B}^\delta(X)$ whose union is "μ almost all of X", in a sense which we now make precise.

A **decomposition for a measure $\boldsymbol{\mu}$** on a δ-ring of $\mathscr{A}$ of subsets of X, or a **$\boldsymbol{\mu}$ decomposition**, is a disjoint subfamily $\mathscr{D}$ of $\mathscr{A}$ such that:

(i) A subset B of $\bigcup \{D : D \in \mathscr{D}\}$ is locally $\mathscr{A}$ measurable if it is **piecewise in** $\mathscr{A}$, in the sense that $B \cap D \in \mathscr{A}$ for each D in $\mathscr{D}$, and

(ii) a member A of $\mathscr{A}$ is of μ measure zero if it is **piecewise of μ measure zero**, in the sense that $\mu(A \cap D) = 0$ for each D in $\mathscr{D}$.

Evidently the union of the members of a μ decomposition is locally $\mathscr{A}$ measurable, and its complement is locally of μ measure zero. A measure μ is **decomposable** iff there is a decomposition for μ.

An example: If $X \in \mathscr{A}_\sigma$ and $\mathscr{D}$ is a disjoint countable subfamily of $\mathscr{A}$ covering X, then $\mathscr{D}$ is a μ decomposition for every measure μ on $\mathscr{A}$, so every measure μ on $\mathscr{A}$ is decomposable.

It is worth noticing that the property of being a decomposition for μ depends only on the families $\mathscr{A}$ and $\mathscr{N} = \{A : A \in \mathscr{A} \text{ and } \mu(A) = 0\}$.

9 PATCHWORK LEMMA *Suppose μ is a measure on a δ-ring $\mathscr{A}$ of subsets of X and $\mathscr{D}$ is a decomposition for μ. Then:*

(i) *if for each member D of $\mathscr{D}$, f_D is an $\mathscr{L}\mathscr{A}$ measurable function with support D, then the function $f = \sum_{D \in \mathscr{D}} f_D$ is $\mathscr{L}\mathscr{A}$ measurable, and,*

(ii) *if ν is a signed measure on $\mathscr{B}$ that is absolutely continuous w.r.t. μ, then $\nu(C) = \sum_{D \in \mathscr{D}} \nu(C \cap D)$ for each C in $\mathscr{B}$.*

PROOF   (i) If $B \in \mathscr{B}(\mathbb{R})$ and $D \in \mathscr{D}$ then $f^{-1}[B] \cap D = f_D^{-1}[B] \cap D \in \mathscr{A}$, so $f^{-1}[B] \cap (\bigcup_{D \in \mathscr{D}} D) \in \mathscr{L}\mathscr{A}$. The set $f^{-1}[B] \cap (X \setminus \bigcup_{D \in \mathscr{D}} D)$ is either empty or $X \setminus \bigcup_{D \in \mathscr{D}} D$ according as B does not or does include zero. Consequently $f^{-1}[B] \in \mathscr{L}\mathscr{A}$ and f is $\mathscr{L}\mathscr{A}$ measurable.

(ii) Since $\nu \prec \mu$, we have $\mathscr{B} \subset \mathscr{A} \subset \mathscr{B}_\sigma$ and (see p. 113) $C \cap A \in \mathscr{B}$ for all A in $\mathscr{A}$. Since $\infty > \mu(C) \geqq \sum_{D \in \mathscr{D}} \mu(C \cap D)$, the family $\mathscr{E} = \{C \cap D : D \in \mathscr{D} \text{ and } \mu(C \cap D) \neq 0\}$ is a countable family of subsets of C, so $\nu(\bigcup_{E \in \mathscr{E}} E) = \sum_{E \in \mathscr{E}} \nu(E)$. The sets $\bigcup \{C \cap D : D \in \mathscr{D} \text{ and } \mu(C \cap D) = 0\}$ and $C \setminus \bigcup \{C \cap D : D \in \mathscr{D}\}$ are piecewise of μ measure zero and hence of measure zero. So $\nu(C) = \sum_{E \in \mathscr{E}} \nu(E) = \sum_{D \in \mathscr{D}} \nu(C \cap D)$. ∎

If for each D in $\mathscr{D}$, ν_D denotes the signed measure defined by $\nu_D(C) = \nu(C \cap D)$ for C in the domain of ν, then the fundamental lemma 5 applied to the positive and negative parts of ν_D yields a μ integrable function f_D with support D such that $\nu_D(C) = (f_D . \mu)(C)$ for all C in the domain of ν. Then the function $f = \sum_{D \in \mathscr{D}} f_D$ is $\mathscr{L}\mathscr{A}$-measurable and $(f . \mu)(C) = \sum_{D \in \mathscr{D}} (f . \mu)(C \cap D) = \sum_{D \in \mathscr{D}} (f_D . \mu)(C) = \sum_{D \in \mathscr{D}} \nu_D(C) = \sum_{D \in \mathscr{D}} \nu(C \cap D) = \nu(C)$ for all C in the domain of ν, according to the preceding lemma. This establishes the following:

10 COROLLARY (RADON–NIKODYM THEOREM FOR DECOMPOSABLE MEASURES) *If μ is a decomposable measure, then each signed measure that is absolutely continuous w.r.t. μ is an indefinite integral w.r.t. μ.*

If $v = f.\mu$, the set $\{x: f(x) > 0\}$ is v positive, and $\{x: f(x) \leqq 0\}$ is v negative, and together they provide a Hahn decomposition of X for v. Thus:

11 COROLLARY (GLOBAL HAHN DECOMPOSITION) *If v is a signed measure and if the variation V_v of v is decomposable, then there is a Hahn decomposition of X for v.*

NOTE It is just as easy to derive the foregoing corollary directly from the patch work lemma: if $\mathscr{D}$ is a V_v decomposition then $\bigcup \{D^+: D \in \mathscr{D}\}$ is v positive and $\bigcup \{D^-: D \in \mathscr{D}\}$ is v negative. ■

Every regular Borel measure μ for a locally compact Hausdorff space X is decomposable, and so the Radon–Nikodym theorem holds for such measures. This will follow immediately from a simple lemma. We agree that a subset K of X is **μ tight** iff it is compact, $\mu(K) > 0$, and $\mu(M) < \mu(K)$ for each compact proper subset M of K. Each compact set A contains a μ tight set K with $\mu(K) = \mu(A)$—explicitly, $K = \{x: x \in A$ *and for each open neighborhood* V *of* x, $\mu(V \cap A) \neq 0\}$. We recall that a set S is a **carrier** for μ iff $S \in \mathscr{L}\mathscr{B}^\delta(X)$ and its complement is locally of μ measure zero.

12 LEMMA *If μ is a regular Borel measure for a locally compact Hausdorff space X and $\mathscr{K}$ is a maximal disjoint family of μ tight sets, then each member of $\mathscr{B}^\delta(X)$ intersects only countably many members of $\mathscr{K}$ and $\bigcup_{K \in \mathscr{K}} K$ is a carrier for μ.*

Moreover, $\mathscr{K}$ is a decomposition for μ.

PROOF We first show that a member V of $\mathscr{B}^\delta(X)$ intersects only countably many members of $\mathscr{K}$. Since V is a subset of an open set in $\mathscr{B}^\delta(X)$ we may assume that V is open. For each positive integer n the family $\{K: K \in \mathscr{K}$ *and* $\mu(K \cap V) \geqq 1/n\}$ is finite since $\mu(V)$ is finite, and for each member K of $\mathscr{K}$ either $\mu(K \cap V) > 0$ or $K \cap V = \varnothing$ because K is μ tight. The assertion follows.

Suppose that for each K in $\mathscr{K}$, $B_K \in \mathscr{B}^\delta(X)$, $B_k \subset K$ and $B = \bigcup_{K \in \mathscr{K}} B_K$. Since a member A of $\mathscr{B}^\delta(X)$ intersects only countably many members of $\mathscr{K}$, $A \cap B$ is the union of countably many members of $\mathscr{B}^\delta(X)$ and is hence a member, and so B is locally Borel.

If $A \in \mathscr{B}^\delta(X)$ then $\{K: K \in \mathscr{K}$ *and* $A \cap K \neq \varnothing\}$ is countable, whence $\mu(A) = \sum_{K \in \mathscr{K}} \mu(A \cap K) + \mu(A \setminus \bigcup_{K \in \mathscr{K}} K)$. If $A \setminus \bigcup_{K \in \mathscr{K}} K$ which belongs to $\mathscr{B}^\delta(X)$ has positive μ measure then it contains a compact set of positive measure and hence a μ tight K of positive measure. This set is clearly disjoint from each member of $\mathscr{K}$ contradicting the maximality of $\mathscr{K}$. Thus $\mathscr{K}$ is a decomposition for μ. ■

13 Corollary (Radon–Nikodym Theorem for Borel Measures) *A signed measure that is absolutely continuous with respect to a regular Borel measure* μ *for a locally compact space is an indefinite integral w.r.t.* μ *of a locally Borel real valued function.*

14 Note The **classical Borel field** for X, $\mathscr{B}_c(X)$ is the σ-field generated by the family of open subsets of X. It is easy to see that $\mathscr{L}\mathscr{B}(X)$ contains $\mathscr{B}_c(X)$, but in general the two fields are not equal. Thus, it can happen that a function f is $\mathscr{L}\mathscr{B}(X)$ measurable but not $\mathscr{B}_c(X)$ measurable. But we do not know whether it is always possible to choose a Radon–Nikodym derivative $d\nu/d\mu$, where μ is a regular Borel measure and $\nu \prec \mu$, to be $\mathscr{B}_c(X)$ measurable.

SUPPLEMENT: HAAR MEASURE

Suppose G is a locally compact Hausdorff topological group. A left Haar measure for G is a regular Borel measure η (a regular measure on $\mathscr{B}^\delta(G)$), not identically zero, that is invariant under left translation. That is, $\eta(z^{-1}B) = \eta(B)$ for all z in G and B in $\mathscr{B}^\delta(G)$, and consequently $\int f(z^{-1}x)\,d\eta x = \int f(x)\,d\eta x$ for each f in $L_1(\eta)$. According to 4.17 there is at least one left Haar measure for G; in fact, to a constant multiple there is precisely one.

We first observe that if η is a left Haar measure for G then $\eta(U) > 0$ for each non-empty open set U in $\mathscr{B}^\delta(G)$ because every compact set is covered by finitely many left translates of U, and hence $\int g\,d\eta > 0$ for each nonnegative member of $C_c(G)$ that is not identically zero. We suppose that g is such a function and show that for each member f of $C_c(G)$, the ratio $\int f\,d\eta/\int g\,d\eta$ is independent of the choice of the left Haar measure η. If ν and η are both left Haar measures for G and $f \in C_c(G)$, then $\int f\,d\eta = k\int f\,d\nu$ with $k = \int g\,d\eta/\int g\,d\nu$, whence, according to the uniqueness part of Riesz theorem 6.16, $\eta = k\nu$. We start with a simple continuity result.

15 Proposition *Suppose* η *is a regular Borel measure for* G *and* $f \in C_c(G)$. *Then the functions* $x \mapsto \int f(ux)\,d\eta u$ *and* $x \mapsto \int f(xv)\,d\eta v$ *are continuous at each member* x *of* G.

PROOF Suppose $x_0 \in G$, W is a compact neighborhood of x_0 and K is a compact support of f. Then for each x in W, the function $u \mapsto f(ux)$ is continuous and has the compact support KW^{-1}. It is straightforward to check that f is left uniformly continuous in the sense that $\sup\{|f(ux) - f(ux_0)|: u \in G\}$ is small for $x_0^{-1}x$ near e, and $|\int f(ux)\,d\eta u - \int f(ux_0)\,d\eta u|$ is also small since $\eta(KW^{-1}) < \infty$. It follows that $x \mapsto \int f(ux)\,d\eta u$ is continuous at x_0 and, in similar fashion, $x \mapsto \int f(xv)\,d\eta v$ is continuous. ∎

The following result is due to v. Neumann.

16 THEOREM (UNIQUENESS OF HAAR MEASURE) *If η and v are left Haar measures for G, then $\eta = kv$ for some positive real number k.*

PROOF Suppose μ and η are left Haar measures for G, g is a non-negative member of $C_c(G)$, not identically zero, and $f \in C_c(G)$. We show that the ratio $\int f\,d\eta/\int g\,d\eta$ is independent of η. According to the preceding proposition, the function $x \mapsto \int g(ux)\,d\mu u$ is continuous and it is strictly positive on G, whence the function h defined on $G \times G$ by $h(x, y) = f(x)g(yx)/\int g(ux)\,d\mu u$ is continuous. Evidently $\int h(x, y)\,d\mu y = f(x)$ for each x in G.

If K is a compact support for f and S for g, then $K \times SK^{-1}$ is a compact support for h, so $h \in C_c(G \times G)$. Consequently by the Tonelli (or Fubini) theorem, $\int h\,d\eta \otimes \mu = \int\int h(x, y)\,d\mu y\,d\eta x = \int f(x)\,d\eta x$. On the other hand, $\int h\,d\eta \otimes \mu = \int\int h(x, y)\,d\eta x\,d\mu y = \int\int h(y^{-1}x, y)\,d\eta x\,d\mu y$ because η is left invariant, $(x, y) \mapsto h(y^{-1}x, y)$ has a compact support and so $\int f(x)\,d\eta x = \int h\,d\eta \otimes \mu = \int\int h(y^{-1}x, y)\,d\mu y\,d\eta x$. Since μ is left invariant, this last integral is $\int\int h(y^{-1}, xy)\,d\mu y\,d\eta x = \int\int f(y^{-1})(g(x)/\int g(uy^{-1})\,d\mu u)\,d\eta x\,d\mu y = \int g\,d\eta \int (f(y^{-1})/\int g(uy^{-1})\,d\mu u)\,d\mu y$. Thus $\int f\,d\eta/\int g\,d\eta$ is independent of η, and so equals $\int f\,dv/\int g\,dv$ for any left Haar measure v, and the theorem follows. ■

For the rest of the supplement **it is assumed that η is a left Haar measure for G.**

If $b \in G$ and $\eta_b(B) = \eta(Bb^{-1})$ for all B in $\mathscr{B}^\delta(G)$, then η_b is a regular, left invariant Borel measure that is not identically zero, and is therefore a real positive multiple of η. The **modular† function Δ** for G is defined by agreeing that $\Delta(b)$, for each b in G, is the unique positive real number such that $\eta(Bb^{-1}) = \Delta(b)\eta(B)$ for all B in $\mathscr{B}^\delta(G)$. Clearly Δ is independent of the choice of left Haar measure and it is easy to see that $\Delta(bc) = \Delta(b)\Delta(c)$. We show that Δ is continuous.

If F is the map $x \mapsto xb$ of G to G and $F\eta$ is Borel image of η under F, then $F\eta(E) = \eta(F^{-1}[E]) = \eta(Eb^{-1}) = \Delta(b)\eta(E)$ for E in $\mathscr{B}^\delta(G)$, and consequently $\int f(xb)\,d\eta x = \Delta(b)\int f(x)\,d\eta x$ for all f in $L_1(\eta)$. Thus right translation $\boldsymbol{R_b}$, where $R_b(f)(x) = f(xb)$, is a linear automorphism of $L_1(\eta)$ which multiplies the norm of each member by $\Delta(b)$. If $f \in C_c(G)$ and $\int f\,d\eta \neq 0$, the relation $\int f(xz)\,d\eta x = \Delta(z)\int f(x)\,d\eta x$ shows (proposition 15) that Δ is continuous. Thus

17 PROPOSITION *Let η be a left Haar measure and let Δ be the modular function for the group G. Then for all a and b in G, E in $\mathscr{B}^\delta(G)$*

† A modular function on a lattice of sets is something quite different (see chapter 1) and "Δ" has already been used for the symmetric difference of two sets. Both usages of "modular" and "Δ" are standard.

and f in $L_1(\eta)$, $\eta(Eb^{-1}) = \Delta(b)\eta(E)$ and $\int f(xb)\,d\eta x = \Delta(b)\int f(x)\,d\eta x$. Moreover, Δ is continuous and $\Delta(ab) = \Delta(a)\Delta(b)$.

It is a consequence of the preceding proposition that if G is compact, then the unimodular function Δ is identically 1, since in this case $\Delta[G]$ is a compact subgroup of the multiplicative group of positive real numbers and $\{1\}$ is the only possibility. The group G is **unimodular** iff the modular function Δ is identically 1. This happens iff each left Haar measure η is also invariant under right translation; that is $\eta(Eb) = \eta(E)$ for all b and all Borel sets E. Compact groups, and of course abelian groups, are unimodular, but even the group of all affine transformations, $x \mapsto ax + b$ with $a \neq 0$ fails to be unimodular.

A **right Haar measure** is a regular Borel measure that is invariant under right translation. If η is left Haar measure and we set $\rho(E) = \eta(E^{-1})$ for all E in $\mathscr{B}(G)$, then ρ is a right Haar measure because $\rho(Eb) = \eta(b^{-1}E^{-1}) = \eta(E^{-1}) = \rho(E)$ for E in $B(G)$ and b in G. The measure **ρ is the right Haar measure corresponding to η**. There is another description of this measure.

18 PROPOSITION *The right Haar measure ρ corresponding to η agrees on $\mathscr{B}^\delta(G)$ with the indefinite integral $\Delta.\eta$.*

Consequently ρ and η are mutually absolutely continuous.

PROOF If $f\Delta$ is η integrable, by the previous result, $\int f(xb)\Delta(xb)\,d\eta x = \Delta(b)\int f(x)\Delta(x)\,d\eta x$ for each b in G. Since $\Delta(xb) = \Delta(x)\Delta(b) \neq 0$, the equation becomes $\int f(xb)\Delta(x)\,d\eta x = \int f(x)\Delta(x)\,d\eta x$, so $\int f(xb)\,d(\Delta.\eta)x = \int f\,d(\Delta.\eta)$. If $B \in \mathscr{B}^\delta(G)$ then $\chi_B\Delta$ is η integrable because Δ is continuous (hence locally $\mathscr{B}^\delta(G)$ measurable) and bounded on the closure of B. Consequently $(\Delta.\eta)|\mathscr{B}^\delta(G)$ is a right invariant Borel measure, and it is regular according to 8.23.

Right Haar measure is essentially unique, like left Haar measure, and it follows that $\rho = k(\Delta.\eta)|\mathscr{B}^\delta(G)$ for some positive number k. We see that $k = 1$ as follows. Choose a neighborhood V of e in $\mathscr{B}^\delta(G)$ so that the continuous function Δ differs very little from 1 on V, and let $W = V \cap V^{-1}$. Then $W = W^{-1}$ and $\rho(W) = \eta(W^{-1}) = \eta(W) > 0$. On the other hand, $\Delta.\eta(W) = \int \chi_W(x)\Delta(x)\,d\eta x$ and since Δ is near 1 on W, $\Delta.\eta(W)$ is near $\eta(W)$, so $k = 1$. ■

19 COROLLARY *If η is a left Haar measure and ρ is the corresponding right Haar measure, then $f \in L_1(\eta)$ if and only if $f/\Delta \in L_1(\rho)$, and in this case $\int f\,d\eta = \int (f/\Delta)\,d\rho = \int f(x^{-1})\Delta(x)\,d\eta$.*

Assume for the rest of this section that G is σ-compact. In this case, if g and h are non-negative members of $L_1(\eta)$, then $(g.\eta) \star (h.\eta) = (g \star_\eta h).\eta$ and $g \star_\eta h \in L_1(\eta)$ according to proposition 8.24.

There is no difficulty in extending the notion of convolution to signed measures and the notion of convolution $\star_\eta$ to functions which are not necessarily non-negative. If the extension is made, then $L_1(\eta)$ becomes an algebra under η and the space $\boldsymbol{M}$ of regular Borel signed measures of finite total variation is a normed algebra under $\star$. Then the preceding conclusion implies that the map $f \mapsto f.\eta$, for f in $L_1(\eta)$ is a norm preserving algebra homomorphism of $L_1(\eta)$ onto the subspace M_a of M which consists of signed measures that are absolutely continuous with respect to η. We notice that a member of M is absolutely continuous w.r.t. η iff it is absolutely continuous w.r.t. some (and hence every) left Haar measure on G, or equivalently w.r.t. some (and hence every) right Haar measure. Thus the space M_a is independent of the choice of the Haar measure.

The η **convolution of functions**, $\star_\eta$, depends on η according to the prescription $(g \star_\eta h)(y) = \int g(x)h(x^{-1}y)\,d\eta x$ for η a.e. y, and replacing η by another left Haar measure changes this convolution by a scalar factor. However $\star_\eta$ is not at all the convolution which is appropriate for members of $L_1(\rho)$, where ρ is the right Haar measure corresponding to η; $\star_\rho$ should be defined so that the map $f \mapsto f.\rho$ for f in $L_1(\rho)$ is a multiplicative map of $L_1(\rho)$ into the algebra of regular signed measures of finite total variation. Since $\rho = \Delta.\eta$, this will be accomplished if the map $f \mapsto (1/\Delta)f$ of $L_1(\rho)$ onto $L_1(\eta)$ is multiplicative; that is,

$$\frac{1}{\Delta}(h \star_\rho f) = \frac{h}{\Delta} \star_\eta \frac{f}{\Delta}$$

for functions h and f belonging to $L_1(\rho)$. A little computation then gives the correct definition for ρ. It is: the **convolution** $h \star_\rho f$ **of functions** h and f belonging to $L_1(\rho)$ is defined by $h \star_\rho f(x) = \int h(xy^{-1})f(y)\,d\rho y$ for ρ a.e. x. We then have the following proposition. The proof is left to the reader.

20 PROPOSITION *Each of the maps shown below is a linear isometry which preserves convolution.*

$$\begin{array}{ccc} L_1(\eta), \star_\eta & \underset{(1/\Delta)f \leftarrow\!\shortmid f}{\overset{f \mapsto \Delta f}{\rightleftarrows}} & L_1(\rho), \star_\rho \\ & {\scriptstyle f \mapsto f.\eta} \searrow \quad \swarrow {\scriptstyle f.\rho \leftarrow\!\shortmid f} & \\ & M, \star & \end{array}$$

Chapter 10

BANACH SPACES

The class of all bounded linear functionals on a normed linear space E is itself a normed linear space E^* called the *dual* or *adjoint* of E. The structure of this space is of interest because a problem about the space E can often be reformulated or "dualized" to a problem about the adjoint space and, if one is lucky, the dual problem may be more amenable to reason than the original. But this dualization usually requires a representation theorem for members of E^*, of the sorts that have already been established (see 6.5 and 6.13). Most of this chapter is devoted to such representations. We begin by reviewing the pertinent definitions.

A **semi-norm** for a real or complex vector space E is a real valued non-negative function $\| \;\|$ on E such that $\|u + v\| \leq \|u\| + \|v\|$ for all u and v in E and $\|ru\| = |r| \, \|u\|$ for u in E and for each scalar r. The semi-norm is a **norm** iff $\|u\| = 0$ only for $u = 0$. The vector space E, with a semi-norm $\| \;\|$ is a **semi-normed** vector space. The semi-metric induced by a semi-norm $\| \;\|$ is given by $dist(u, v) = \|u - v\|$ for all u and v in E. If the semi-norm $\| \;\|$ is in fact a norm, the induced semi-metric is a metric and if, further, this metric space is complete, then E with $\| \;\|$, is a **Banach space**.

Here are some classical examples and some classical notation. The space $\boldsymbol{m}$ of all bounded sequences $s = \{s_n\}_n$ of real numbers, with the supremum norm $\|s\|_{\mathbb{N}} = sup_{n \in \mathbb{N}} |s_n|$, is a normed space. It is identical with $L_\infty(\gamma)$, where γ is counting measure for $\mathbb{N}$, with the norm $\| \;\|_\infty$, and it is frequently denoted ℓ_∞. According to theorem 6.10, if μ is a measure, then $L_\infty(\mu)$, with $\| \;\|_\infty$, is complete and consequently ℓ_∞ is a Banach space.

The space c consisting of all convergent sequences of real numbers, with the supremum norm, is a closed subspace of m. It is closed because the uniform limit of continuous functions is continuous and (roughly speaking) c is the set of functions on $\mathbb{N}$ that can be extended continuously to $\mathbb{N} \cup \{+\infty\}$. Thus c is a Banach space, and so is the subspace c_0 consisting of all sequences of real numbers that converge to zero. We will presently describe the adjoint of the space c_0.

A linear functional ϕ on a space E with semi-norm $\| \; \|$ is continuous w.r.t. the semi-metric induced by $\| \; \|$ iff it is **bounded**, in the sense that $\|\phi\| = sup\{|\phi(u)|: u \in E \text{ and } \|u\| \leqq 1\} < \infty$. In this case $|\phi(u)| \leqq \|\phi\| \, \|u\|$—indeed $\|\phi\| = inf\{r: |\phi(u)| \leqq r\|u\| \text{ for all } u \text{ in } E\}$—hence $\|\phi(u) - \phi(v)\| \leqq \|\phi\| \, \|u - v\|$ for all u and v, and in particular ϕ is uniformly continuous.

The class of all bounded linear functionals on a semi-normed space E is the **dual** or **adjoint** or **conjugate** E^* of E. It is normed by the **dual norm** $\phi \mapsto \|\phi\|$, which is just the supremum norm on the unit ball $B = \{u: u \in E \text{ and } \|u\| \leqq 1\}$. Convergence in the metric induced by this norm is uniform convergence in the complete normed space of all bounded continuous functions on B with the supremum norm. Consequently E^* is complete whether or not E is complete, and so E^* is a Banach space.

Here is an example of a representation theorem for an adjoint space. Suppose $\phi \in c_0^*$, that for k and m in $\mathbb{N}$, $\delta^k(m) = 1$ if $k = m$ and zero otherwise, and that $f(k) = \phi(\delta^k)$. For u in c_0, the supremum norm $\|u - \sum_{k=1}^n u_k \delta^k\|_{\mathbb{N}}$ converges to zero because u vanishes at ∞ and, since ϕ is continuous, $\phi(u) = lim_n \sum_{k=1}^n u_k \phi(\delta^k) = lim_n \sum_{k=1}^n u_k f(k)$. Then $\phi(signum\, f) = lim_n \sum_{k=1}^n |f(k)| = \sum_k |f(k)| < \infty$. Consequently $f \in L_1(\gamma)$, $\phi(u) = \int u_k f(k)\, d\gamma\, k$, $|\phi(u)| \leqq \|u\|_{\mathbb{N}} \|f\|_1$ and $\phi(signum\, f) = \|f\|_1$ so $\|\phi\| = \|f\|_1$. Hence $f \mapsto (u \mapsto \int uf \, d\gamma)$ is a linear isometry of $L_1(\gamma)$ onto c_0^*. The space $L_1(\gamma)$ is also denoted ℓ_1, and one says (inaccurately) that ℓ_1 is the adjoint of c_0, $\ell_1 = (c_0)^*$.

There is a suggestive way of phrasing this last result. A signed measure μ on the δ-ring $\mathscr{A}$ of finite subsets of $\mathbb{N}$ is the indefinite integral $f.\gamma$ for some f in $L_1(\gamma) = \ell_1$ iff it is of finite total variation, so c_0^* can be identified as the space of all signed measures on $\mathscr{A}$ that are of finite total variation.

Let $\boldsymbol{C_0(\mathbb{R})}$ be the class of all continuous real valued functions on $\mathbb{R}$ that **vanish at** ∞ in the sense that for each $e > 0$, the set $\{x: |f(x)| \geqq e\}$ is bounded. We will identify $C_0(\mathbb{R})^*$ after a technical lemma.

Recall that the **variation** $\boldsymbol{V_\mu}$ of a signed measure μ on A is $\mu^+ + \mu^-$, and the **total variation of** $\boldsymbol{\mu}$, $\|\mu\|_V$, is $sup\{V_\mu(A): A \in \mathscr{A}\}$. Alternatively, $\|\mu\|_V = sup\{\mu(A) - \mu(B): A \text{ and } B \text{ disjoint members of } \mathscr{A}\} = sup\{\sum_{i=1}^k |\mu(A_i)|: \{A_i\}_{i=1}^k \text{ a disjoint family in } \mathscr{A}\}$.

1 Lemma *Suppose μ is a signed measure on a δ-ring $\mathscr{A}$ of subsets of X, $\|\mu\|_V < \infty$, and S is the class of $\mathscr{A}$ σ-simple functions on X with finite supremum norm $\| \;\|_X$.*

Then each member f of S is μ integrable, $|\int f\,d\mu| \leqq \|f\|_X \|\mu\|_V$, and if ϕ is the linear functional $f \mapsto \int f\,d\mu$, then $\|\phi\| = \|\mu\|_V$.

PROOF Each member f of S has a support E in $\mathscr{A}_\sigma$, consequently $|f| \leqq \|f\|_X \chi_E$, so f is V_μ integrable and $|\int f\,dV_\mu| \leqq \|f\|_X \int \chi_E\,dV_\mu \leqq \|f\|_X \|\mu\|_V$, so $\|\phi\| \leqq \|\mu\|_V$. On the other hand, each member A of $\mathscr{A}$ is the union of a μ positive set A^+ and a μ negative set A^-, and if $f = \chi_{A^+} - \chi_{A^-}$ then $\phi(f) = V_\mu(A) = \|f\|_X V_\mu(A)$, so $\|\phi\| \geqq V_\mu(A)$ for A in $\mathscr{A}$, hence $\|\phi\| \geqq \|\mu\|_V$ and equality results. ■

We recall that $\boldsymbol{C_c(\mathbb{R})}$ is the family of all continuous real valued functions on $\mathbb{R}$ that have compact support.

2 Riesz Representation Theorem *For each bounded linear functional ϕ on $C_0(\mathbb{R})$ there is a unique signed Borel measure μ for $\mathbb{R}$ such that $\phi(f) = \int f\,d\mu$ for all f in $C_0(\mathbb{R})$. Moreover, $\|\phi\| = \|\mu\|_V$.*

PROOF The space $C_0(\mathbb{R})$ with its natural ordering is a vector lattice and consequently ϕ is the difference $\phi^+ - \phi^-$ of positive linear functionals provided $\sup\{\phi(u)\colon 0 \leqq u \leqq f\} < \infty$ for all $f \geqq 0$ (see chapter 0). But if $0 \leqq u \leqq f$, then $\|u\|_{\mathbb{R}} \leqq \|f\|_{\mathbb{R}}$ and so $|\phi(u)| \leqq \|\phi\|\,\|u\|_{\mathbb{R}} \leqq \|\phi\|\,\|f\|_{\mathbb{R}} < \infty$. Hence $\phi = \phi^+ - \phi^-$ for some positive linear functionals ϕ^+ and ϕ^- on $C_0(\mathbb{R})$.

According to theorem 6.5, each positive linear functional on $C_c(\mathbb{R})$ is equal on $C_c(\mathbb{R})$ to the integral w.r.t. a Borel measure μ for $\mathbb{R}$, and consequently $\phi(f)$ is the integral of f w.r.t. a signed Borel measure μ for $\mathbb{R}$ for each f in $C_c(\mathbb{R})$. The preceding lemma then implies that each member f of $C_c(\mathbb{R})$ is μ integrable, $|\int f\,d\mu| \leqq \|f\|_X \|\mu\|_V$ and $\sup\{|\phi(f)|\colon f \in C_c(\mathbb{R}) \text{ and } \|f\|_X \leqq 1\} = \|\mu\|_V$. But $C_c(\mathbb{R})$ is dense in $C_0(\mathbb{R})$. It follows that this supremum is just $\|\phi\|$, and $\phi(f) = \int f\,d\mu$ for f in $C_0(\mathbb{R})$. The uniqueness of μ follows from the equality of norms. ■

3 Note The same proof establishes the corresponding result for the space $C_{[a:b]}$ of continuous real valued functions on $[a:b]$, with the supremum norm (a **Borel measure for $[a:b]$** is a measure on the δ-ring generated by compact subsets of $[a:b]$).

We have established a generalization of the proposition that $c_0{}^*$ is (essentially) ℓ_1, and it is natural to seek a description of $(c_0{}^*)^* = \ell_1{}^*$. This is not hard. If ϕ is a bounded linear functional on ℓ_1 and $g_n =$

$\phi(\delta^n)$ for all n in $\mathbb{N}$, then g is a bounded sequence and $\phi(s) = \sum_n s_n g_n$ for all s in ℓ_1. On the other hand, if g is an arbitrary member of ℓ_∞, then $s \mapsto \sum_n s_n g_n$ is a member ϕ_g of ℓ_1, and $\|\phi_g\|$ is $\|g\|_\infty$. Consequently $\ell_1^* = \{\phi_g : g \in \ell_\infty\}$. Since $\ell_1 = L_1(\gamma)$ and $\ell_\infty = L_\infty(\gamma)$, where γ is counting measure for $\mathbb{N}$, this result suggests a representation theorem for $L_1(\mu)^*$ for an arbitrary measure μ. We establish such a theorem as well as results for $L_p{}^*$ for all p, $1 \leqq p \leqq \infty$.

Suppose $1 \leqq p \leqq \infty$, q is the **index conjugate to p** (that is, $1/p + 1/q = 1$), and $g \in L_q(\mu)$. Then $gf \in L_1(\mu)$ and $\|gf\|_1 \leqq \|g\|_q \|f\|_p$ for all f in $L_p(\mu)$, according to the Hölder inequality 6.7. Consequently, if $\boldsymbol{\phi_g}$ is defined by $\phi_g(f) = \int gf\,d\mu$ for f in $L_p(\mu)$, then $\phi_g \in L_p(\mu)^*$ and $\|\phi_g\| \leqq \|g\|_q$. This last inequality is in fact an equality. A stronger form of this result is established below.

4 LEMMA *Suppose μ is a measure on $\mathscr{A}$, $1 \leqq p \leqq \infty$, q is the index conjugate to p, g is an $\mathscr{A}$ σ-simple function such that $gf \in L_1(\mu)$ for all $\mathscr{A}$ simple functions f and $\sup\{|\int gf\,d\mu| : f$ $\mathscr{A}$ simple and $\|f\|_p = 1\} < \infty$.*

Then $g \in L_q(\mu)$ and $\|\phi_g\| = \|g\|_q$.

PROOF Let *signum* g be the sign of g, so $(\mathit{signum}\,g)\,g = |g|$ and let $M_q(g) = \sup\{|\int gf\,d\mu| : f$ $\mathscr{A}$ simple *and* $\|f\|_p = 1\}$. We prove that $\|g\|_q \leqq M_q(g)$ whence $\|g\|_q$ is finite and equals $\|\phi_g\|$. For $p = 1$, so $q = \infty$, we prove that for $e > 0$, the set $S = \{x : |g(x)| \geqq M_\infty(g) + e\}$ is locally of μ measure zero, so $\|g\|_\infty \leqq M_q(g)$. If $A \in \mathscr{A}$ and $A \subset S$, then $M_\infty(g)\mu(A) \geqq \int(\chi_A\,\mathit{signum}\,g)g\,d\mu = \int_A |g|\,d\mu \geqq (M_\infty(g) + e)\mu(A)$. So $0 \geqq e\mu(A)$ and $\mu(A) = 0$.

Suppose $1 < p \leqq \infty$, so q is finite. Let $\{A_n\}_n$ be an increasing sequence of members of $\mathscr{A}$ whose union is a support for g and let $\{g_n\}_n$ be a sequence of $\mathscr{A}$ simple functions dominated by $|g|$ so that g_n has support A_n for each n and $\{g_n\}_n$ converges to g pointwise. Let $h_n = (\mathit{signum}\,g)(|g_n|/\|g_n\|_q)^{q-1}$. Then $h_n \in L_p(\mu)$ and $\|h_n\|_p = 1$, and by Fatou's lemma, $\|g\|_q \leqq \liminf_n \|g_n\|_q = \liminf_n \int |g_n h_n|\,d\mu \leqq \liminf_n \int |g h_n|\,d\mu = \liminf_n \int g h_n\,d\mu \leqq M_q(g)$. ■

5 LEMMA *Suppose μ is a measure on $\mathscr{A}$ and $1 \leqq p \leqq \infty$. Then:*

(i) *The space $L_p(\mu)^*$ is a vector lattice.*
(ii) *If $\phi \in L_p(\mu)^*$ and $v(A) = \phi(\chi_A)$ for A in $\mathscr{A}$, then v is finitely additive and if $p < \infty$, then v is countably additive.*
(iii) *If $p < \infty$, ϕ is a positive member of $L_p(\mu)^*$, $E \in \mathscr{A}$, and $v_E(A) = \phi(\chi_{A \cap E})$ for all A in $\mathscr{A}$, then there is a non-negative member g_E of $L_1(\mu)$ with support E such that $v_E = g_E \cdot \mu$.*

PROOF To prove (i) it is sufficient, in view of the decomposition lemma of chapter 0, to prove that if $f \in L_p(\mu)$ and $\phi \in L_p(\mu)^*$, then

$sup\{\phi(u): u \in L_p(\mu)$ and $0 \leqq u \leqq f\} < \infty$. But if $0 \leqq u \leqq f$ then $\|u\|_p \leqq \|f\|_p$ and $|\phi(u)| \leqq \|\phi\| \, \|u\|_p \leqq \|\phi\| \, \|f_p\| < \infty$.

The first assertion of (ii), that ν is finitely additive, follows from the fact that the functional ϕ is additive. If a member A of $\mathscr{A}$ is the union of an increasing sequence $\{A_n\}_n$ in $\mathscr{A}$, then $lim_n \|\chi_A - \chi_{A_n}\|_p = 0$, provided $p < \infty$, and so $\nu(A) = \phi(\chi_A) = lim_n \phi(\chi_{A_n}) = lim_n \nu(A_n)$. Consequently, if $p < \infty$ then ν is countably additive.

The third assertion follows directly from the fundamental lemma 9.5 that was used to establish the Radon–Nikodym theorem. According to 9.5, $\nu_E(A) = \int ((1 - h)/h) \, d\mu$ for A in $\mathscr{A}$ that is a subset of $E \backslash E_0$ where E_0 is a set of μ measure zero. But then E_0 is of ν_E measure zero and consequently, changing h on a μ-null set, we have $\nu_E = g_E . \mu$ for a member g_E of $L_1(\mu)$. ■

6 Representation Theorem *Suppose μ is a measure on a δ-ring $\mathscr{A}$ of subsets of X, p and q are conjugate indices, $1 \leqq p < \infty$ and for each g in $L_q(\mu)$, $\phi_g(f) = \int fg \, d\mu$ for all f in $L_p(\mu)$.*

If $p > 1$, or if $p = 1$ and $X \in \mathscr{A}_\sigma$, then each member ϕ of $L_p(\mu)^*$ is ϕ_g for some g in $L_q(\mu)$ and $\|\phi\| = \|g\|_q$.

PROOF We show that a member ϕ of $L_p(\mu)^*$ is ϕ_g for some g in $L_q(\mu)$. Because $L_p(\mu)^*$ is a vector lattice, ϕ is the difference of positive members of $L_p(\mu)^*$ and so we may assume that ϕ is a positive linear functional. In this case the correspondence $A \mapsto \phi(\chi_A)$ is a measure ν on $\mathscr{A}$, and if $E \in \mathscr{A}$ and $\nu_E(A) = \nu(A \cap E)$ then, according to lemma 3, $\nu_E = g_E . \mu$ for some μ integrable function g_E with support E. We obtain the required function g by "piecing together the functions g_E".

If $B \in \mathscr{A}$ and $p > 1$, then $\nu(B) = \phi(\chi_B) \leqq \|\phi\| \, \|\chi_B\| = \|\phi\| \mu(B)$ and consequently $sup_{B \in \mathscr{A}} \nu(B) < \infty$. Hence there is an increasing sequence $\{B_n\}_n$ in $\mathscr{A}$ such that $sup_{B \in \mathscr{A}} \nu(B) = lim_n \nu(B_n)$, and therefore the complement of $\bigcup_n B_n$ is locally of ν measure zero. If $E_n = B_{n+1} \backslash B_n$ for each n, then $\nu(A) = \sum_n \nu(A \cap E_n) = \sum_n \nu_{E_n}(A) = \sum g_{E_n} . \mu(A) = g.\mu(A)$ for all A in $\mathscr{A}$, where $g = \sum_n g_{E_n}$. If $p = 1$ and $X \in \mathscr{A}_\sigma$, then $X = \bigcup_n E_n$ for some disjoint sequence in $\mathscr{A}$ and $\nu(A) = \sum_n \nu(A \cap E_n)$ and we again conclude that $\nu(A) = g.\mu(A)$. Thus $\phi(\chi_A) = \int \chi_A g \, d\mu$ for all A in $\mathscr{A}$ whence $\phi(f) = \int fg \, d\mu$ for simple functions f. According to lemma 4, $g \in L_q(\mu)$ and $\|g\|_q = \|\phi_g\|$, and since ϕ and ϕ_g agree on the simple functions in $L_p(\mu)$, $\phi(f) = \phi_g(f)$ for all f. ■

It is worth noticing that if $1 \leqq p < \infty$ then a member ϕ of $L_p(\mu)^*$ determines, up to a member of $L_q(\mu)$ of norm zero, the function g such that $\phi = \phi_g$. That is, if h is another such function then $\|g - h\|_q = 0$ since $0 = \phi_g - \phi_h = \phi_{g-h}$ and $\|g - h\|_q = \|\phi_{g-h}\|$. Here is another way of describing the preceding theorem: the map $g \mapsto \phi_g$ for $g \in L_q(\mu)$ is, for $1 \leqq p < \infty$, a norm preserving linear map of $L_q(\mu)$ onto $L_p(\mu)^*$.

The preceding theorem does not hold if $p = \infty$—it is not the case that $\ell_\infty{}^*$ is ℓ_1—but there is a reasonable sounding description of the adjoint of $L_\infty(\mu)$ for an arbitrary measure μ. Suppose that μ is a measure on $\mathscr{A}$ and $\mathscr{L}\mathscr{A}$ is the σ-field of locally $\mathscr{A}$ measurable sets. Then the $\mathscr{L}\mathscr{A}$ simple real valued functions are dense in $L_\infty(\mu)$ according to 6.10. Consequently each bounded linear functional ϕ on $L_\infty(\mu)$ is the unique continuous extension of $\phi|\{f: f \text{ is } \mathscr{L}\mathscr{A} \text{ simple}\}$. We are thus led to seek a representation for the adjoint of the space of $\mathscr{B}$ simple functions on X, with the supremum norm $\| \ \|_X$, where $\mathscr{B}$ is an arbitrary ring of subsets of X.

Let us agree, for convenience, that if v is real valued and additive on a ring $\mathscr{B}$ of sets, then $\|v\|_V = sup\{|\mu(A) - \mu(B)|: A \text{ and } B \text{ disjoint members of } \mathscr{B}\}$. If $\mathscr{B}$ is a δ-ring and v is a signed measure, then this definition agrees with the earlier usage of $\| \ \|_V$. Let $L^{\mathscr{B}}$ be the space of $\mathscr{B}$ simple functions, and for each linear functional ϕ on $L^{\mathscr{B}}$, let $\|\phi\| = sup\{|\phi(f)|: f \in L^{\mathscr{B}} \text{ and } \|f\|_X \leqq 1\}$, whether or not the supremum is finite. Then $|\phi(f)| \leqq \|f\|_X \|\phi\|$ for all f in $L^{\mathscr{B}}$ if we agree that $0 \cdot \infty = 0$.

7 LEMMA *If $\mathscr{B}$ is a ring of subsets of X, ϕ is a linear functional on $L^{\mathscr{B}}$ and $v(B) = \phi(\chi_B)$ for B in $\mathscr{B}$, then $\|\phi\| = \|v\|_V$ and $|\phi(f)| \leqq \|f\|_X \|v\|_V$.*

PROOF Each $\mathscr{B}$ simple function f on X is $\mathscr{C}$ simple for some finite subring $\mathscr{C}$ of $\mathscr{B}$. Then $\mathscr{C}$ is a δ-ring, $v|\mathscr{C}$ is a signed measure of finite total variation and $\phi(f) = \int f \, dv|\mathscr{C}$, so lemma 1 implies that $|\phi(f)| \leqq \|f\|_X \|v|\mathscr{C}\|_V$ and $sup\{|\phi(f)|: f \in L^{\mathscr{C}} \text{ and } \|f\|_X \leqq 1\} = \|v|\mathscr{C}\|_V$. It is possible to choose a finite subring $\mathscr{C}$ of $\mathscr{B}$ so that $sup\{|\phi(f)|: f \in L^{\mathscr{C}} \text{ and } \|f\|_X \leqq 1\}$ is near $\|\phi\|$ and $\|v|\mathscr{C}\|_V$ is near $\|v\|_V$. The desired result follows. ■

For each ring $\mathscr{B}$ of subsets of X, $L^{\mathscr{B}}$ is the space of $\mathscr{B}$ simple functions, with the supremum norm $\| \ \|_X$, $\boldsymbol{FA(\mathscr{B})}$ is defined to be $\{v: v: \mathscr{B} \to \mathbb{R}, v \text{ is finitely additive and } \|v\|_V < \infty\}$, with the norm $\| \ \|_V$, and, if $\mathscr{B}$ is a δ-ring, $\boldsymbol{M(\mathscr{B})}$ is the subspace of $FA(\mathscr{B})$ consisting of signed measures. Lastly, for each linear functional ϕ on a vector space E of real valued functions on X, v_ϕ is defined by $v_\phi(A) = \phi(\chi_A)$ for all A such that $\chi_A \in E$.

8 THEOREM *If $\mathscr{B}$ is a ring of subsets of X and if for each ϕ in $(L^{\mathscr{B}})^*$, $v_\phi(B) = \phi(\chi_B)$ for B in $\mathscr{B}$, then the map $\phi \mapsto v_\phi$ is a linear isometry of $(L^{\mathscr{B}})^*$ onto $FA(\mathscr{B})$, and in particular $FA(\mathscr{B})$ is complete.*

If $\mathscr{B}$ is a δ-ring, then $M(\mathscr{B})$ is a closed subspace of $FA(\mathscr{B})$ and is therefore complete.

PROOF To prove the first assertion we need only show that each member of $FA(\mathscr{B})$ is v_ϕ for a member ϕ in $(L^{\mathscr{B}})^*$. If $f \in L^{\mathscr{B}}$ then $f = \sum_{i=1}^n b_i \chi_{B_i}$ for some real numbers b_i and some disjoint members B_i of $\mathscr{B}$, $i = 1, 2, \ldots, n$. For $v \in FA(\mathscr{B})$, let I^v be the unique linear extension of the correspondence $\chi_B \mapsto v(B)$ for B in $\mathscr{B}$ (see 2.6). Then $|I^v(f)| = |\sum_{i=1}^n b_i v(B_i)| \leqq (max_i |b_i|) \sum_{i=1}^n |v(B_i)| \leqq \|f\|_X \|v\|_V$ whence I^v is a bounded linear functional ϕ and evidently $v_\phi = v$.

Each bounded linear functional ϕ on $L^{\mathscr{B}}$ is the difference $\phi^+ - \phi^-$ of positive linear functionals because $(L^{\mathscr{B}})^*$ is a vector lattice and the decomposition lemma of chapter 0 applies. Consequently each member μ of $FA(\mathscr{B})$ is the difference $\mu^+ - \mu^-$ of positive members. Let $M = \{v\colon v \in FA(\mathscr{B})$ and $lim_n\, v(B_n) = v(\prod_n B_n)$ for each decreasing sequence $\{B_n\}_n$ in $\mathscr{B}$. Evidently M is closed (the uniform limit of continuous functions is continuous). If μ is a signed measure, both μ^+ and μ^- belong to M, so $\mu \in M$. On the other hand if $\mu \in M$ whence μ^+ and μ^- both belong to M, they are both measures and so μ is a signed measure. Thus $M = M(\mathscr{B})$. ■

Suppose μ is a measure on a δ-ring $\mathscr{A}$ of subsets of X, $\mathscr{L}\mathscr{A}$ is the σ-field of locally $\mathscr{A}$ measurable sets and $\mathscr{N}$ is the subfamily of $\mathscr{L}\mathscr{A}$ consisting of sets that are locally of measure zero. If ϕ is a linear functional on $L^{\mathscr{L}\mathscr{A}}$ and $v(A) = \phi(\chi_A)$ for A in $\mathscr{L}\mathscr{A}$, then $\|v\|_V = sup\{|\phi(f)|\colon f \in L^{\mathscr{L}\mathscr{A}}$ and $\|f\|_X \leqq 1\}$ and, if v vanishes on $\mathscr{N}$, this is the same as $sup\{|\phi(f)|\colon f \in L^{\mathscr{L}\mathscr{A}}$ and $\|f\|_\infty \leqq 1\}$. If $\phi \in L_\infty(\mu)^*$ whence v vanishes on $\mathscr{N}$, the last supremum is $\|\phi\|$ because $L^{\mathscr{L}\mathscr{A}}$ is dense in $L_\infty(\mu)$ according to 6.10. It follows that

9 THEOREM *Suppose μ is a measure on $\mathscr{A}$, $\mathscr{N}$ is the family of $\mathscr{L}\mathscr{A}$ measurable sets that are locally of μ measure 0, $\phi \in L_\infty(\mu)^*$ and $v_\phi(A) = \phi(\chi_A)$.*

Then v_ϕ is finitely additive on $\mathscr{L}\mathscr{A}$, vanishes on $\mathscr{N}$ and $\|v_\phi\|_V = \|\phi\|$, and every finitely additive function on $\mathscr{L}\mathscr{A}$ that vanishes on $\mathscr{N}$ is v_ϕ for some ϕ in $L_\infty(\mu)^*$.

10 DIGRESSION The semi-normed spaces E that we have encountered are almost all of a special type. They are **semi-normed vector lattices**, in the sense that E is a vector lattice with a norm $\|\ \|$ such that $\||x|\| = \|x\|$ for all $x \in E$, and $\|x\| \geqq \|y\|$ provided $x \geqq y \geqq 0$.

Such a space E is an **L space**, or of **type L_1**, iff $\|x + y\| = \|x\| + \|y\|$ for all positive x and y in E. The spaces c_0^*, ℓ_1, $L_1(\mu)$, $C_0(\mathbb{R})^*$, $FA(\mathscr{B})$ for a ring $\mathscr{B}$ of sets, and $M(\mathscr{A})$ for a δ-ring $\mathscr{A}$, are all L-spaces.

A semi-normed lattice E is an M **space** iff $\|x \vee y\| = \max\{\|x\|, \|y\|\}$ for all positive members x and y of E. The spaces c, m, ℓ_∞, $C_0(\mathbb{R})$, $L_\infty(\mu)$ and $L^{\mathscr{B}}$ are M spaces.

In each example that we have seen, the adjoint of an L space is an M space and the adjoint of an M space is an L space. S. Kakutani, to whom these concepts are due, has developed a structure theory for L spaces. (See, for example, the appendix to [KN]).

A semi-normed vector lattice E is of **type p**, or an **L_p space**, $1 < p < \infty$, iff $\|x\|^p + \|y\|^p = \|x + y\|^p$ for all positive members x and y of E. The adjoint of such a space is of type q, where q is the index conjugate to p. (See F. Bohnenblust, *Duke Math. J.* **6** (1940), 627–640.)

All of the examples at our disposal indicate that the **second adjoint**, or **double dual**, E^{**} of a normed space E is like E, or at least that E is like a subspace of E^{**}. For example, if $1 < p < \infty$ and μ is a measure, then $L_p(\mu)^*$ is like $L_q(\mu)$, and $L_p(\mu)^{**}$ is like $L_p(\mu)$. But the second adjoint E^{**} may "contain" E as a proper subspace. For example, c_0^{**} is essentially ℓ_1^* which is like ℓ_∞, and $c_0 \subset \ell_\infty$ but $c_0 \neq \ell_\infty$.

If E is an arbitrary semi-normed space and $x \in E$, then $f \mapsto f(x)$, for f in E^*, is a linear functional on E^* that is called **evaluation $\mathscr{E}_x$ at x**. Formally, for each member x of E, $\mathscr{E}_x(f) = f(x)$ for all f in E^*. Evidently $|\mathscr{E}_x(f)| \leqq \|x\| \, \|f\|$, so $\mathscr{E}_x \in (E^*)^*$ and $\|\mathscr{E}_x\| \leqq \|x\|$. Thus $\mathscr{E}$ is a linear map, **evaluation**, of E into E^{**}, and it is **bounded** in the sense that $\|\mathscr{E}\| = sup\{\|\mathscr{E}_x\|: x \in E \text{ and } \|x\| \leqq 1\} < \infty$. It is the case, but we do not prove till later, that $\mathscr{E}$ is an isometry of E into the second adjoint E^{**}.

The evaluation map $\mathscr{E}$ may carry E *onto* E^{**}, in which case E is said to be **reflexive**, or the image $\mathscr{E}[E]$ may be a proper subspace of E^{**} so that E is **non-reflexive**. There are at hand examples of both sorts of spaces. It seems reasonable, and is in fact the case, that if μ is a measure and $1 < p < \infty$, then $L_p(\mu)$ is reflexive. It is also to be expected (in view of 6.13) that each Hilbert space is reflexive. On the other hand, it is unlikely that c_0, ℓ_1, ℓ_∞ or $C_0(\mathbb{R})$ is reflexive.

SUPPLEMENT: THE SPACES $C_0(X)^*$ AND $L_1(\mu)^*$

The Riesz representation theorem 2 for $C_0(\mathbb{R})^*$ has a straightforward generalization to $C_0(X)$, for a locally compact Hausdorff space X. According to theorem 6.16, each positive linear functional ϕ on $C_0(X)$, is of the form $f \mapsto \int f\,dv$ for a unique regular Borel measure v for X with finite total variation $\|v\|_V$, and $\|\phi\| = \|v\|_V$. On the other hand, each bounded linear functional on $C_0(X)$ is the difference of two positive linear functionals, according to proposition 2.13. This establishes the first part of the following theorem. We agree that v is a **regular Borel signed measure** iff it is the difference of regular Borel measures.

11 RIESZ REPRESENTATION THEOREM *Each bounded linear functional ϕ on $C_0(X)$ agrees on $C_0(X)$ with the integral w.r.t. a regular Borel signed measure v of finite total variation. Moreover, v is unique and $\|\phi\| = \|v\|_V$.*

PROOF It is only the equality of norms that must be proved. The uniqueness then follows for if v and μ are regular Borel signed measures of finite total variation and $\phi(f) = \int f\,dv = \int f\,d\mu$ for all f in $C_0(X)$, then $v - \mu$ is a regular Borel signed measure which represents the zero functional Z on $C_0(X)$, whence $\|v - \mu\|_V = \|Z\| = 0$.

Suppose then that $\phi(f) = \int f\,dv$ for all f in $C_0(X)$. For $e > 0$ there are disjoint compact sets A and B such that $|v(A) - v(B) - \|v\|_V| < e$ by regularity. If $C \in \mathscr{B}^\delta(X)$ and is a subset of $X \backslash (A \cup B)$, then $|v(C)| < e$. There is a member f of $C_0(X)$ so that f is 1 on A, -1 on B and $\|f\|_X = 1$. Then $\int f\,dv$ differs from $\int (\chi_A - \chi_B)\,dv$ by at most e, and hence $\int f\,dv$ differs from $\|v\|_V$ by at most $2e$. Consequently $\|v\|_V = sup\{|\int f\,dv| : f \in C_0(X)\ and\ \|f\|_X = 1\}$. ■

The representation theorem 6 for $L_1(\mu)^*$ extends to decomposable measures.

12 REPRESENTATION THEOREM FOR $L_1(\mu)^*$ *If μ is a decomposable measure and $\phi \in L_1(\mu)^*$, then there is a member g of $L_\infty(\mu)$ such that $\phi(f) = \int gf\,d\mu$ for all f in $L_1(\mu)$. The function g is determined locally μ a.e. by ϕ and $\|\phi\| = \|\eta\|_\infty$.*

PROOF According to the representation theorem 6, for each A in the domain $\mathscr{A}$ of μ, the functional ϕ_A on $L_1(\mu)$ defined by $\phi_A(f) = \phi(\chi_A f)$, is represented by a member g_A of $L_\infty(\mu)$ with support A, so $\phi(\chi_A f) = \int g_A f\,d\mu$ for all f in $L_1(\mu)$. By modifying g_A on a set of μ measure zero we may assume that $\|g_A\|_X = \|\phi_A\| \leqq \|\phi\|$. If $\mathscr{D}$ is a decomposition for μ, the function $g = \sum_{D \in \mathscr{D}} g_D$ is $\mathscr{L}\mathscr{A}$ measurable according to the patchwork lemma 9, and $\|g\|_X \leqq \|\phi\|$.

We complete the proof by establishing that $\phi(f) = \int gf\,d\mu$ for f in $L_1(\mu)$. Each member A of $\mathscr{A}$ intersects only countably many members of $\mathscr{D}$ in a set of positive measure (because $\infty > \mu(A) \geqq \sum_{D \in \mathscr{D}} \mu(A \cap D)$) and each f in $L_1(\mu)$ has support in $\mathscr{A}_\sigma$, whence there is a countable subfamily $\mathscr{E}$ of $\mathscr{D}$ such that $f(x) = 0$ for almost all x in $X \backslash \bigcup_{D \in \mathscr{E}} D$ and $\{\chi_D f\}_{D \in \mathscr{E}}$ is pointwise summable μ a.e. and norm summable to f in $L_1(\mu)$. Consequently $\phi(f) = \sum_{D \in \mathscr{E}} \phi(\chi_D f) = \sum_{D \in \mathscr{E}} \int g_D f\,d\mu = \int (\sum_{D \in \mathscr{E}} g_D) f\,d\mu = \int gf\,d\mu$. ■

SUPPLEMENT: COMPLEX INTEGRAL AND COMPLEX MEASURE

Suppose μ is a measure on a δ-ring $\mathscr{A}$ of subsets of X. A complex valued $f: X \to \mathbb{C}$ is **integrable w.r.t. μ** and $f \in \boldsymbol{L_1(\mu:\mathbb{C})}$ iff there are sequences $\{A_n\}_n$ in $\mathscr{A}$ and $\{a_n\}_n$ in $\mathbb{C}$ such that $\sum_n |a_n| \mu(A_n) < \infty$ and $f(x) = \sum_n a_n \chi_{A_n}(x)$ for all x in X, and in this case $\boldsymbol{\int f\,d\mu = I_\mu(f)} = \sum_n a_n \mu(A_n)$. This definition is not ambiguous. The function f is integrable w.r.t. μ iff $\mathscr{R}f$ and $\mathscr{I}f$ are integrable and in this case $\int f\,d\mu = \int \mathscr{R}f\,d\mu + i\int \mathscr{I}f\,d\mu$. The space of complex valued integrable functions is a vector space over $\mathbb{C}$ and $f \mapsto \int f\,d\mu$ is a complex linear functional on it. If it is necessary to make a distinction between the spaces of complex valued integrable and real valued integrable functions we will denote the spaces respectively as $L_1(\mu:\mathbb{C})$ and $L_1(\mu:\mathbb{R})$; if the context makes the underlying scalar field clear, we shall use $L_1(\mu)$ to denote, ambigously, either space. If f is μ integrable, so is $|\mu|$ and $|\int f\,d\mu| \leqq \int |f|\,d\mu = \|f\|_1$, $\|\ \|_1$ is a semi-norm, and $L_1(\mu:\mathbb{C})$ is complete under the corresponding semi-metric. The spaces $L_p(\mu:\mathbb{C})$ and the p-norms $\|\ \|_p$ $(1 < p \leqq \infty)$ are defined as in the real case and the completeness theorem extends to $L_p(\mu:\mathbb{C})$ for all p. The inner product for $L_2(\mu:\mathbb{C})$ is given by $\langle f, g\rangle = \int fg^{\sim}\,d\mu$ where $g^{\sim}$ is the complex conjugate of g.

There is a useful extension of the *signum* function. If $x \in \mathbb{R}$, then $signum\, x = 0$ if $x = 0$, 1 if $x > 0$ and -1 if $x < 0$, whence $signum\, x = x/|x|$ for $x \neq 0$. For each complex number x, if $x = 0$, then $signum\, x = 0$, and if $x \neq 0$, then $signum\ x = x/|x|$. Thus $signum\ x$ is a complex number of modulus one if $x \neq 0$, and $x(signum\, x)^{\sim} = |x|$ for all x. (It is customary to write $signum\, x$ instead of $signum(x)$, and if g is a complex valued function $(signum\, g)(u) = signum\, g(u)$ for all u.)

The **real restriction** $E_{\mathbb{R}}$ of a complex vector space E is the same set E with the same vector addition, and with scalar multiplication restricted to $\mathbb{R} \times E$. In brief, $E_{\mathbb{R}}$ is just E if you forget about scalar multiplication by imaginary numbers. If ϕ is a linear functional on E, then the real part $\mathscr{R}\phi$ of ϕ is a linear functional on $E_{\mathbb{R}}$ and $\mathscr{R}\phi$ determines ϕ, because $\phi(x) = \mathscr{R}\phi(x) - i\mathscr{R}\phi(ix)$ for each x in E. If ψ is an arbitrary linear functional on $E_{\mathbb{R}}$, then $x \mapsto \psi(x) - i\psi(ix)$, for x in E, is a linear functional on E (notice that $ix \mapsto \psi(ix) - i\psi(-x) = i(\psi(x) - i\psi(ix))$. There is thus a one-to-one correspondence between linear functionals on E and linear functionals on $E_{\mathbb{R}}$. If E is a semi-normed space and if ϕ is a bounded linear functional on E, then $sup\{|\phi(x)|: \|x\| \leqq 1\} = sup\{|\mathscr{R}\phi(x)|: \|x\| \leqq 1\}$. This is true because $|\phi(x)| = \phi(x/signum\,\phi(x))$. Consequently the correspondence $\phi \mapsto \mathscr{R}\phi$ is an isometric mapping of E^* onto $E_{\mathbb{R}}{}^*$. In fact it is a linear isometry of $(E^*)_{\mathbb{R}}$ onto $(E_{\mathbb{R}})^*$. Thus:

If E is a semi-normed complex vector space, then the map $\phi \mapsto \mathscr{R}\phi$ for ϕ in E^* is an $\mathbb{R}$-linear isometry of $(E^*)_{\mathbb{R}}$ onto $E_{\mathbb{R}}{}^*$.

The theorems and proofs for $L_p(\mu:\mathbb{R})$ which do not involve the order relation for $\mathbb{R}$ (e.g., the dominated convergence theorem and the Hölder and the Minkowski inequalities) carry over to $L_p(\mu:\mathbb{C})$. In particular, if p and q are conjugate indices, and $g \in L_q(\mu:\mathbb{C})$, then the complex linear functional ϕ_g on $L_p(\mu:\mathbb{C})$ defined by $\phi_g(f) = \int fg^{\sim}\,d\mu$ is bounded, $|\phi_g(f)| \leqq \|g\|_q \|f\|_p$ and $\|\phi_g\| \leqq \|g\|_q$. As in the real case, this inequality is actually an equality, and so $g \mapsto \phi_g$ is a linear isometry of $L_q(\mu:\mathbb{C})$

onto $L_p(\mu:\mathbb{C})^*$ if $1 < p < \infty$, and also for $p = 1$ if $X \in \mathscr{A}_\sigma$ or, more generally, if μ is a decomposable measure.

Suppose μ is a measure on a δ-ring $\mathscr{A}$ for X and f is a complex valued $\mathscr{L}\mathscr{A}$ measurable function on X (that is, f is $\mathscr{L}\mathscr{A} - \mathscr{B}(\mathbb{R}^2)$ measurable, which is the case iff $\mathscr{R}f$ and $\mathscr{I}f$ are $\mathscr{L}\mathscr{A}$ measurable). The **indefinite integral** of f w.r.t. μ, $\boldsymbol{f.\mu}$, is defined by $\boldsymbol{f.\mu}(B) = \int_B f\,d\mu$ for each member B of $\mathscr{B} = \{B\colon B \in \mathscr{A}$ *and* $f\chi_B \in L_1(\mu)\}$. Then $\mathscr{B}$ is a δ-ring and $\mathscr{B} \subset \mathscr{A} \subset \mathscr{B}_\sigma$. If $\{B_n\}_n$ is a disjoint sequence in $\mathscr{B}$ such that $\bigcup_n B_n \in \mathscr{B}$, then $f.\mu(B) = \sum_n f.\mu(B_n)$, so $f.\mu$ is a **complex valued measure**, or just a **complex measure** on $\mathscr{B}$. If η is a complex measure on a δ-ring $\mathscr{B}$, then $\mathscr{R}\lambda$ and $\mathscr{I}\lambda$ are signed measures on $\mathscr{B}$ and if ρ and η are any signed measures on $\mathscr{B}$, then $\rho + i\eta$ is a complex measure on $\mathscr{B}$.

The **variation of a complex measure** λ on $\mathscr{B}$ is defined by letting $V_\lambda(B)$, for B in $\mathscr{B}$, equal $sup \sum_{A \in \mathscr{C}} |\lambda(A)|$ for all finite subfamilies $\mathscr{C}$ of $\mathscr{B}$ that consist of disjoint subsets of B. If one changes this definition by replacing "finite" by countable, then the same function V_λ results, and this makes it easy to verify that V_λ is a measure. The **total variation** of λ, $\|\lambda\|_V$, is $sup\{V_\lambda(B)\colon B \in \mathscr{B}\}$, and the class of complex measures λ for which $\|\lambda\|_V < \infty$ is a complex vector space that is normed by $\| \;\; \|_V$.

A complex measure λ on a δ-ring $\mathscr{B}$ for X is **absolutely continuous w.r.t. a measure $\boldsymbol{\mu}$** on a δ-ring $\mathscr{A}$ for X, and we write $\boldsymbol{\lambda \prec \mu}$, iff $V_\lambda \prec \mu$. That is, $\mathscr{B} \subset \mathscr{A} \subset \mathscr{B}_\sigma$ and $V_\lambda(B) = 0$ for all members B of $\mathscr{B}$ for which $\mu(B) = 0$. This is the case iff $\mathscr{R}\lambda$ and $\mathscr{I}\lambda$ are absolutely continuous w.r.t. μ. If $X \in \mathscr{A}_\sigma$, or more generally, if there is a decomposition for μ, then λ is an indefinite integral w.r.t. μ, so $\lambda = f.\mu|\mathscr{B}$ for some $\mathscr{L}\mathscr{A}$ measurable complex function f on X. The function f, which is determined up to a set that is locally of μ measure zero, is a **Radon–Nikodym derivative of the complex measure $\boldsymbol{\lambda}$** w.r.t. μ, denoted $\boldsymbol{d\lambda/d\mu}$.

Suppose that μ on $\mathscr{A}$ is a measure and f is an $\mathscr{L}\mathscr{A}$ measurable function. If f is real valued so that $f.\mu$ is a signed measure, we have seen that $V_{f.\mu} = |f|.\mu$. The conclusion extends to the complex case because: Suppose that f is complex valued, and B is an arbitrary member of the domain $\mathscr{B}$ of $f.\mu$. For each disjoint family $\{B_i\}_{i=1}^k$ of members of $\mathscr{B}$ that are subsets of B, $\sum_{i=1}^k |(f.\mu)(B_i)| = \sum_{i=1}^k |\int_{B_i} f\,d\mu| \leqq \sum_{i=1}^k \int_{B_i} |f|\,d\mu \leqq \int_B |f|\,d\mu$ whence $V_{f.\mu}(B) \leqq (|f|.\mu)(B)$. On the other hand if $\sum_{i=1}^n b_i\chi_{B_i}$ is a $\mathscr{B}$ simple function that is uniformly close to the function $\chi_B(signum\, f)$ and bounded by 1 in absolute value (proposition 6.10), then $(|f|.\mu)(B)$ which is $\int f\chi_B(signum\, f)\,d\mu$, is close to $\int f(\sum_{i=1}^k b_i\chi_{B_i})\,d\mu$ whose absolute value is at most $V_{f.\mu}(B)$. Consequently $V_{f.\mu}(B) = (|f|.\mu)(B)$ for all B in $\mathscr{B}$, i.e., $V_{f.\mu} = |f|.\mu$.

It is evident that each complex measure λ is absolutely continuous w.r.t. V_λ and hence $\lambda = g.V_\lambda$ for some g, provided V_λ is decomposable. In this case $V_\lambda = |g|.V_\lambda$ and consequently $|g| = 1$ locally V_λ almost everywhere. The decomposition $\lambda = g.V_\lambda$ with $|g| = 1$ is called the **polar decomposition of** λ.

The **product complex measure** $\lambda_1 \otimes \lambda_2$ of complex measures λ_1 on $\mathscr{A}$ and λ_2 on $\mathscr{B}$ agrees on $A \times B$ for A in $\mathscr{A}$ and B in $\mathscr{B}$ with $\lambda_1(A)\lambda_2(B)$. If $\lambda_1 = g.V_{\lambda_1}$ and $\lambda_2 = h.V_{\lambda_2}$ then $\lambda_1(A)\lambda_2(B) = \int_{A\times B} g(x)h(y)\,dV_{\lambda_1} \otimes V_{\lambda_2}$. This makes a Fubini theorem for $\lambda_1 \otimes \lambda_2$ derivable from that for $V_{\lambda_1} \otimes V_{\lambda_2}$.

A complex measure on the Borel δ-ring $\mathscr{B}\delta(X)$ of a locally compact Hausdorff space X, is a **complex Borel measure for** X. It is **regular** iff the measure V_λ is regular, and this is the case iff the signed measures $\mathscr{R}\lambda$ and $\mathscr{I}\lambda$, or equivalently their positive and negative parts, are regular. The family $M_r(\mathscr{B}^\delta(X):\mathbb{C})$ of all regular complex Borel measures is a closed subspace, under the total variation norm, of the Banach space of all complex Borel measures for X and is therefore a Banach space (if λ is regular, $\|\nu - \lambda\|_V < e$ and $V_\lambda(U\backslash A) < e$, then $V_\nu(U\backslash A) \leqq \|\nu - \lambda\|_V + V_\lambda(U\backslash A) < 2e$).

Suppose $C_0(X:\mathbb{C})$ is the Banach space of complex valued continuous functions on X that vanish at ∞, with the supremum norm $\|\ \|_X$. For each regular complex Borel measure $\lambda = \rho + i\eta$, the functional ϕ_λ on $C_0(X:\mathbb{C})$ defined by $\phi_\lambda(f) = \int f\,d\rho + i\int f\,d\eta$, is a member of $C_0(X:\mathbb{C})^*$, and the correspondence $\lambda \mapsto \phi_\lambda$ is a linear isometry of $M_r(\mathscr{B}^\delta(X):\mathbb{C})$ onto $C_0(X:\mathbb{C})^*$.

Each Borel measure is decomposable and consequently each complex Borel measure λ for X is an indefinite integral $f.\mu$ for some Borel measure μ for X and for some locally $\mathscr{A}$ measurable f. If the measure μ is regular then so is $f.\mu$ and so is $|f|.\mu$. Moreover, the correspondence $f \mapsto f.\mu$ is a linear isometry of $L_1(\mu:\mathbb{C})$ onto the space of all regular complex Borel measures for X that are of finite variation and are absolutely continuous w.r.t. μ, with the total variation norm (see 8.23).

SUPPLEMENT: THE BOCHNER INTEGRAL

The construction of an integral for real or complex valued functions generalizes in a natural way to an integration process for vector valued functions. If μ is a (non-negative finite valued) measure on a δ-ring of subsets of X, then to certain functions on X with values in a Banach space E, we assign a member of E, denoted $\int f(x)\,d\mu x$. This assignment, $f \mapsto \int f\,d\mu$ is the *Bochner integral.* The basic properties of the integral of scalar valued functions generalize to the Bochner integral, except for those (e.g., Fatou's lemma) which depend explicitly on the ordering of the real numbers.

If f is an E valued function on X and φ is a bounded linear functional on E, then the composition $\varphi \circ f$ is a scalar valued function on X. It will turn out that if $f\colon X \to E$ has Bochner integral $\int f\,d\mu$

and $\varphi \in E^*$, then $\varphi(\int f\,d\mu) = \int \varphi \circ f\,d\mu$. This connection between the Bochner integral and the scalar integral is very useful, and for this and other reasons we need an ample supply of bounded linear functionals on E. The Hahn–Banach extension theorem shows that there is such a supply.

13 HAHN–BANACH EXTENSION THEOREM *Suppose F is a subspace of a real vector space E and that p is a non-negative real valued function on E such that $p(x + y) \leqq p(x) + p(y)$ and $p(tx) = tp(x)$ for all x and y in E and t in $\mathbb{R}$.*

Then each linear functional φ on F with $\varphi \leqq p$ extends to a linear functional φ' on E with $\varphi' \leqq p$.

PROOF Let P be the class of all linear functionals ψ on subspaces of E such that $\psi \leqq p$ and ψ is an extension of φ. The class P is ordered by agreeing that ψ follows θ iff ψ is an extension of θ. Each linearly ordered subset of P has an upper bound (its graph is the union of the graphs of its members), and consequently, by the Hausdorff principle, there is a maximal member of P. We need to prove that the domain F of a maximal member is all of E and this will follow from maximality if we show that if $x \in E\backslash F$, then it is possible to extend a linear functional dominated by p on F to the larger space $\{rx + y : r \in \mathbb{R}, y \in F\}$, so that the extension is dominated by p.

If there is a real number t such that $rt + \varphi(y) \leqq p(rx + y)$ for all real numbers r and all y in F, then $\varphi'(rx + y) = rt + \varphi(y)$ defines the desired extension. If $r > 0$, then $rt + \varphi(y) \leqq p(rx + y)$ iff $t \leqq p(x + y/r) - \varphi(y/r)$. That is, t must be less than or equal to $p(x + u) - \varphi(u)$ for all u in F. If $r < 0$, then the inequality reduces in the same way to $-p(-x + v) + \varphi(v) \leqq t$ for all v in F. Consequently there is a number t as required unless $-p(-x + v) + \varphi(v) > p(x + u) - \varphi(u)$ for some members u and v of F. But in this case $\varphi(u + v) > p(-x + v) + p(x + u) \geqq p(u + v)$, which is a contradiction. ■

14 COROLLARY *If F is a subspace of a semi-normed space E and $\varphi \in F^*$, then there is an extension ψ of φ in E^* such that $\|\varphi\| = \|\psi\|$. Consequently for each y in E if $y \neq 0$, then there is ψ in E^* so that $\|\psi\| = \psi(y/\|y\|) = 1$.*

PROOF If E is a real linear space, then the corollary is a special case of the preceding theorem with $p(x) = \|\varphi\|\,\|x\|$ for all x in E. If E is a complex normed space then $\mathcal{R}\varphi$ is a real linear functional on the real restriction $F_{\mathbb{R}}$ of F and it is easy to see that $\|\mathcal{R}\varphi\| = \|\varphi\|$. Consequently there is a real linear functional θ on $E_{\mathbb{R}}$ so that $\|\theta\| = \|\mathcal{R}\varphi\|$ and θ is an extension of $\mathbb{R}\varphi$. Then, if $\psi(x) = \theta(x) - i\theta(ix)$ for x in E, ψ is a complex linear functional on E, (notice that $\psi(ix) = \theta(ix) - i\theta(-x) =$

$i(\theta(x) - i\theta(ix)) = i\psi(x))$, and $\|\psi\| = \|\theta\| = \|\mathscr{R}\varphi\| = \|\varphi\|$. Finally, ψ is an extension of φ because $\mathscr{R}\psi$ is an extension of $\mathscr{R}\varphi$, so $\mathscr{R}(\psi - \varphi)$ vanishes on F, and hence $\psi = \varphi$ on F.

If $x \in E$ and $x \neq 0$, the foregoing conclusion applied to the linear functional φ defined by $\varphi(\alpha y) = \alpha \| y \|$ on the space F of scalar multiples of y yields the last statement of the corollary. ■

We recall that for each x in E, the evaluation at x, $\mathscr{E}_x$, is the functional on E^* defined by $\mathscr{E}_x(f) = f(x)$ for f in E^*. It is evident that $|\mathscr{E}_x(f)| \leqq \|x\| \, \|f\|$ whence $\|\mathscr{E}_x\| \leqq \|x\|$. This inequality is actually an equality because the preceding corollary guarantees the existence of a member f in E^* of norm one with $\mathscr{E}_x(f) = \|x\|$. Thus

15 COROLLARY *The evaluation map of a semi-normed space E into its second adjoint E^{**} is a linear isometry.*

We now define measurability and then Bochner integrability for E valued functions. **We assume throughout that E is a normed linear space over either $\mathbb{R}$ or $\mathbb{C}$, that E^* is the dual of E, and that μ is a (finite valued) measure on a δ-ring $\mathscr{A}$ of subsets of X.**

The **strong Borel σ-field** for E, or the **classical Borel σ-field, $\mathscr{B}_c(E)$**, is the σ-field generated by the family of open subsets of E. The members of $\mathscr{B}_c(E)$ are **strongly Borel measurable**, or **strongly Borel**. If E is separable (there is a countable dense subset of E), then $\mathscr{B}_c(E)$ is generated by any base for the topology, since each open set is the union of countably many members of the base. The **strong Borel field for a subset A of E** is the σ-field $\mathscr{B}_c\|A = \{B \cap A : B \in \mathscr{B}_c(E)\}$. It is generated by $\{S \cap A : S \in \mathscr{S}\}$ where $\mathscr{S}$ is any family generating $\mathscr{B}_c(E)$.

The weak Borel field $\mathscr{W}$, or **$\mathscr{W}(E)$**, is the smallest σ-field such that every bounded linear functional φ on E is measurable. The strong Borel field $\mathscr{B}_c$ is one such σ-field, and so $\mathscr{W}$ is a subfamily of $\mathscr{B}_c$. The **weak Borel field for a subset A of E** is the relativization $\mathscr{W}\|A = \{W \cap A : W \in \mathscr{W}\}$.

The σ-fields $\mathscr{B}_c$ and $\mathscr{W}$ are generally not identical. However:

16 THEOREM *If A is a separable subset of E, then $\mathscr{B}_c\|A = \mathscr{W}\|A$.*

PROOF We first show that there is a sequence $\{\varphi_n\}_n$ in E^* such that $\|x - y\| = sup_n |\varphi_n(x - y)|$ for all x and y in A. To this end, select a dense sequence $\{z_n\}_n$ in A, and for all nonnegative integers i and j choose a linear functional φ_{ij} of norm one such that $\|z_i - z_j\| = \varphi_{ij}(z_i - z_j)$. This choice is possible because of the Hahn-Banach theorem. Let $p(x) = sup_{i,j} |\varphi_{ij}(x)|$ for all members x of E. Then $p(x) \leqq \|x\|$ because $\|\varphi_{ij}\| = 1$ for all i and j, and $p(z_i - z_j) = \|z_i - z_j\|$. It is easy to verify that p is a semi-norm. We show that $p(x - y) = \|x - y\|$ for all members x and y of A by an approximation argument.

Choose z_i and z_j so that $\|x - z_i\|$ and $\|y - z_j\|$ are small. Then, from the triangle inequality, $\|x - y\|$ is near $\|z_i - z_j\|$. But $p(x - z_i)$ and $p(y - z_j)$ are also small because $p(u) \leqq \|u\|$ for all u, and from the triangle inequality (p is a semi-norm), we again have $p(x - y)$ near $p(z_i - z_j) = \|z_i - z_j\|$. Consequently $p(x - y)$ is near $\|x - y\|$.

The preceding result shows that the function $y \mapsto \|x - y\|$, for y in A, is $\mathscr{W}\|A$ measurable because it is the supremum of a countable family of $\mathscr{W}\|A$ measurable functions. Consequently, for every such x and every $r > 0$ the set $\{y\colon y \in A \text{ and } \|x - y\| < r\}$ is $\mathscr{W}\|A$ measurable. But sets of this form generate $\mathscr{B}_c\|A$, so $\mathscr{B}_c\|A \subset \mathscr{W}\|A$. ■

We recall that if $\mathscr{A}$ is a δ-ring of subsets of X, then $\mathscr{L}\mathscr{A}$ is the family of all subsets of X which are locally in $\mathscr{A}$, in the sense that $B \cap A \in \mathscr{A}$ for all members A of $\mathscr{A}$. The family $\mathscr{L}\mathscr{A}$ is a σ-field for X. We agree that a function f on X to E is **strongly measurable**, iff f is $\mathscr{L}\mathscr{A} - \mathscr{B}_c$ measurable; that is, $f^{-1}[S]$ is locally $\mathscr{A}$ measurable, for each member S of the strong Borel field. The function f is **weakly measurable** iff f is $\mathscr{L}\mathscr{A} - \mathscr{W}$ measurable. The family of all sets of the form $\varphi^{-1}[B]$, with φ in E^* and B a Borel subset of the scalars, generates $\mathscr{W}$, so f is weakly measurable iff $f^{-1}[\varphi^{-1}[B]] = (\varphi \circ f)^{-1}[B] \in \mathscr{L}\mathscr{A}$ for all such φ and B. That is, f is weakly measurable iff $\varphi \circ f$ is $\mathscr{L}\mathscr{A}$ measurable for all φ in E^*.

Each strongly measurable function f on X to E is automatically weakly measurable. If a function f on X to E is weakly measurable *and* $f[X]$ is separable, then for each strong Borel set B, $f^{-1}[B] = f^{-1}[B \cap f[X]] \in \mathscr{B}_c\|f[X]$ according to theorem 16, and hence f is strongly measurable. Thus:

17 COROLLARY *If $f\colon X \to E$ is weakly measurable and has separable range, then f is strongly measurable.*

The description of weak measurability in terms of measurability of the functions $\varphi \circ f$, with φ in E^*, makes it clear that the sum of two weakly measurable E valued functions and a scalar multiple of each such function are weakly measurable. The pointwise limit f of a sequence $\{f_n\}_n$ of weakly measurable functions on X to E is weakly measurable because, for each φ in E^*, $\varphi \circ f$ is the pointwise limit of the sequence $\{\varphi \circ f_n\}_n$ of measurable scalar valued functions and is therefore measurable. On the other hand, the class of strongly measurable functions is not necessarily a linear space, although closed under pointwise sequential convergence. But we are concerned only with functions with separable range, and the preceding corollary guarantees that there are no surprises here.

The relation between measurable E valued functions and σ-simple functions is much the same as in the scalar case. A function f on X to E is $\mathscr{A}$ **simple** iff there are *finite* sequences $\{a_k\}_k$ in E and $\{A_k\}_k$ in $\mathscr{A}$ such that $f(x) = \sum_k \chi_{A_k}(x) a_k$ for all x in X, and a function f on X to E is $\mathscr{A}$ **σ-simple** or just **σ-simple** iff there are sequences $\{a_n\}_n$ in E and $\{A_n\}_n$ in $\mathscr{A}$ such that $f(x) = \sum_n \chi_{A_n}(x) a_n$ for x in X. To be precise: it is required that for each x, the unordered sum of the sequence $\{\chi_{A_n}(x) a_n\}_n$ relative to the norm topology be $f(x)$.

Each σ-simple function $f = \sum_n \chi_{A_n} a_n$ has a support $\bigcup_n A_n$ in $\mathscr{A}_\sigma$ and a separable range--indeed the set of linear combinations with rational coefficients of finitely many members of $\{a_n\}_n$ is dense in a separable subspace of E that contains $f[X]$. Moreover, a σ-simple function f is, being the pointwise limit of a sequence of $\mathscr{A}$ simple functions, weakly and therefore strongly measurable. Conversely, it is the case that each strongly measurable E valued function f with separable range and a support in $\mathscr{A}_\sigma$ is σ-simple. We prove a stronger statement which gives information on both f and $\|f\|$.

If f is an $\mathscr{L}\mathscr{A}$ measurable *real* valued function with support in $\mathscr{A}_\sigma$, then by theorem 5.9 f^+ and f^- are countable linear combinations with nonnegative coefficients of characteristic functions of members of $\mathscr{A}$, so there are sequences $\{r_n\}_n$ in $\mathbb{R}$ and $\{A_n\}_n$ in $\mathscr{A}$ so that $f = \sum_n r_n \chi_{A_n}$ and $|f| = \sum_n |r_n| \chi_{A_n}$. Essentially the same conclusion holds for E valued functions f, except that the second equality is replaced by an epsilon close inequality. We prove this after establishing a preliminary lemma. An E valued function f is $\mathscr{A}$ **elementary** iff for some sequence $\{a_n\}_n$ in E and some disjoint sequence $\{A_n\}_n$ in $\mathscr{A}$, $f(x) = \sum_n \chi_{A_n}(x) a_n$ for each x in X.

18 LEMMA *Each strongly measurable E valued function f with separable range and a support in $\mathscr{A}_\sigma$ is the uniform limit of a sequence of E valued $\mathscr{A}$ elementary functions.*

PROOF Suppose that $\bigcup_n A_n$, where $\{A_n\}_n$ is a disjoint sequence in $\mathscr{A}$, is a support for f and that for each n, $\{y_{n,m}\}_m$ is a dense subset of $f[A_n]$. For $e > 0$ let $E_{n,m} = \{x: x \in A_n \text{ and } \|f(x) - y_{n,m}\| < e\}$. Then $E_{n,m} \in \mathscr{A}$ because f is strongly measurable, and "disjointing" by setting $F_{n,m} = E_{n,m} \setminus \bigcup_{k<m} E_{n,k}$, we see that $\{F_{n,m}\}_{n,m}$ is a disjoint countable subfamily of $\mathscr{A}$. If $g(x) = y_{n,m}$ for x in $F_{n,m}$, and $g(x) = 0$ for x outside the support $\bigcup_n A_n$ of f, then g is elementary and $\|g(x) - f(x)\| < e$ for all x. ■

19 REPRESENTATION THEOREM *Suppose that $f: X \to E$ is strongly measurable, has separable range and a support $\bigcup_k C_k$ for some disjoint sequence $\{C_k\}_k$ in $\mathscr{A}$.*

Then for each sequence $\{e_k\}_k$ of positive numbers, there are sequences

$\{A_n\}_n$ in $\mathscr{A}$ and $\{a_n\}_n$ in E such that $f(x) = \sum_n \chi_{A_n}(x) a_n$ and $\|f(x)\| \leqq \sum_n \chi_{A_n}(x)\|a_n\| \leqq \|f(x)\| + \sum_k e_k \chi_{C_k}(x)$ for all x.

PROOF It is sufficient to show that if f is supported by a member C of $\mathscr{A}$ and $e > 0$ then $f(x) = \sum_n \chi_{A_n}(x) a_n$ and $\sum_n \chi_{A_n}(x)\|a_n\| \leqq \|f(x)\| + e\chi_C(x)$ for some sequences $\{a_n\}_n$ in E and $\{A_n\}_n$ in $\mathscr{A}$ with $A_n \subset C$ for each n.

We suppose f is supported by a member C of $\mathscr{A}$. The function f is the uniform limit of a sequence of elementary functions, according to lemma 18 and consequently, by using the differencing trick, we can find a sequence $\{f_n\}_n$ of elementary functions, each supported by C, such that $f(x) = \sum_n f_n(x)$ for each x and $\|f_n(x)\| < e2^{-n}$ for all x and for $n > 0$. Then $\|f_0(x)\| \leqq \|f(x)\| + \sum_{n=1}^{\infty} e2^{-n} = \|f(x)\| + e$ for each x. If $f_n(x) = \sum_k \chi_{B_{n,k}}(x) b_{n,k}$, where $\{B_{n,k}\}_k$ is a disjoint sequence in $\mathscr{A}$ with $B_{n,k} \subset C$, then $f(x) = \sum_{n,k} \chi_{B_{n,k}}(x) b_{n,k}$ for all x. For each n, because f_n is elementary, $\|f_n(x)\| = \sum_k \chi_{B_{n,k}}(x)\|b_{n,k}\|$, whence $\sum_{n,k} \chi_{B_{n,k}}(x)\|b_{n,k}\| = \sum_n \|f_n(x)\| \leqq \|f_0(x)\| + e \leqq \|f(x)\| + 2e$. Since C is a support for f, $\|f(x)\| \leqq \sum_{n,k} \chi_{B_{n,k}}(x)\|b_{n,k}\| \leqq \|f(x)\| + 2e\chi_C$. This establishes the lemma. ■

20 PROPOSITION *For each function f on X to E the following are equivalent:*

(i) *f is weakly measurable, has separable range and a support in $\mathscr{A}_\sigma$,*
(ii) *f is strongly measurable, has separable range and a support in $\mathscr{A}_\sigma$,*
(iii) *f is $\mathscr{A}$ σ-simple, and*
(iv) *f is the pointwise limit of a sequence of $\mathscr{A}$ simple functions.*

PROOF Corollary 5 shows that (i) implies (ii), the preceding theorem shows that (ii) implies (iii), and (iii) clearly implies (iv). Finally, (iv) implies (i) follows from the fact that the pointwise limit f of a sequence $\{f_n\}_n$ of weakly measurable functions on X to E is weakly measurable. ■

A function f on X to E is **Bochner integrable** iff there are sequences $\{A_n\}_n$ in $\mathscr{A}$ and $\{a_n\}_n$ in E with $\sum_n \mu(A_n)\|a_n\| < \infty$ such that $f(x) = \sum_n \chi_{A_n}(x) a_n$ for each x. We want to define the Bochner integral of such a function f to be $\sum_n \mu(A_n) a_n$. If E is complete the sequence $\{\mu(A_n) a_n\}_n$ is summable to a member of E, because it is absolutely summable. We **assume henceforth that E is complete** and we show that the desired definition is not ambiguous.

21 FUNDAMENTAL LEMMA *If f is an E valued function, $f(x) = \sum_n \chi_{A_n}(x) a_n$ for each x, and $\sum_n \mu(A_n)\|a_n\| < \infty$, then $\varphi \circ f \in L_1(\mu)$ for each φ in E^* and $\int \varphi \circ f \, d\mu = \varphi(\sum_n \mu(A_n) a_n$.*

If $f(x)$ is also equal to $\sum_n \chi_{B_n}(x) b_n$ for all x, where $\sum_n \mu(B_n) \|b_n\| < \infty$, then $\sum_n \mu(A_n) a_n = \sum_n \mu(B_n) b_n$.

PROOF Clearly $\varphi \circ f(x) = \sum_n \chi_{A_n}(x) \varphi(a_n)$ and $\sum_n \mu(A_n) |\varphi(a_n)| \leqq \|\varphi\| \sum_n \mu(A_n) \|a_n\| < \infty$. Consequently $\varphi \circ f \in L_1(\mu)$ and $\int \varphi \circ f \, d\mu = \sum_n \mu(A_n) \varphi(a_n) = \varphi(\sum_n \mu(A_n) a_n)$. This establishes the first statement of the lemma. The second statement follows from the Hahn–Banach theorem (corollary 14): since $\varphi(\sum_n \mu(A_n) a_n - \sum_n \mu(B_n) b_n) = 0$ for all φ in E^*, $\sum_n \mu(A_n) a_n - \sum_n \mu(B_n) b_n = 0$. ∎

If f is a Bochner integrable function on X to a Banach space E, $f(x) = \sum_n \chi_{A_n}(x) a_n$ for each x, and $\sum_n \mu(A_n) \|a_n\| < \infty$, then the **Bochner integral** of f, $\int f \, d\mu$ or $B_\mu(f)$, is defined to be $\sum_n \mu(A_n) a_n$. The preceding lemma shows that this definition is not ambiguous. It also shows that if f is Bochner integrable, then $\varphi \circ f$ is integrable for every φ in E^* and $\int \varphi \circ f \, d\mu = \varphi(\int f \, d\mu)$. We notice that $\|\sum_n \chi_{A_n}(x) a_n\| \leqq \sum_n \chi_{A_n}(x)(x) \|a_n\|$ for each x, and it follows that if f is Bochner integrable then $x \mapsto \|f(x)\|$ is integrable.

We show that conversely, if f is an $\mathscr{A}$ σ-simple function and $x \mapsto \|f(x)\|$ is integrable, then f is Bochner integrable and moreover, $\|B_\mu(f)\| \leqq \int \|f(x)\| \, d\mu x$.

22 THEOREM *If f is an $\mathscr{A}$ σ-simple function on X to E then f is Bochner integrable if and only if $x \mapsto \|f(x)\|$ is integrable, and in this case, $\|\int f \, d\mu\| \leqq \int \|f\| \, d\mu$. Moreover, if T is a bounded linear map of E to a Banach space F then $T \circ f$ is Bochner integrable and $T(\int f \, d\mu) = \int T \circ f \, d\mu$.*

PROOF We have already established some of the assertions of the theorem. It only remains to show that if $x \mapsto \|f(x)\|$ is integrable, then so is f and $\|\int f \, d\mu\| \leqq \int \|f\| \, d\mu$. According to theorem 19, for each sequence $\{e_k\}_k$ of positive numbers, $f(x) = \sum_n \chi_{A_n}(x) a_n$ and $\sum_n \chi_{A_n}(x) \|a_n\| \leqq \|f(x)\| + \sum_k e_k \chi_{C_k}(x)$ for all x for some sequences $\{A_n\}_n$ in $\mathscr{A}$ and $\{a_n\}_n$ in E and some disjoint sequence $\{C_k\}_k$ in $\mathscr{A}$ so that $\bigcup_k C_k$ is a support for f. Let $e_k = e 2^{-k} / \mu(C_k)$ for each k for which $\mu(C_k) \neq 0$ and 0 otherwise. Then the σ-simple real valued function $x \mapsto \sum_n \chi_{A_n}(x) \|a_n\|$ is dominated by $x \mapsto \|f(x)\| + \sum_k e_k \chi_{C_k}(x)$ which is integrable. Consequently $x \mapsto \sum_n \chi_{A_n}(x) \|a_n\|$ is integrable and its integral $\sum_n \mu(A_n) \|a_n\| \leqq \int \|f\| \, d\mu + e$. So f is Bochner integrable and $\|\int f \, d\mu\| \leqq \sum_n \mu(A_n) \|a_n\| \leqq \int \|f\| \, d\mu + e$ whence $\|\int f \, d\mu\| \leqq \int \|f\| \, d\mu$. ∎

For each Banach space E and each measure μ, the class of Bochner integrable functions is denoted by $L_1(\mu, E)$. For f in $L_1(\mu, E)$, $\|f\|_1$ is defined to be $\int \|f(x)\| \, d\mu x$. It is easy to see that $L_1(\mu, E)$ is a linear space and that $\| \ \|_1$ is a semi-norm for $L_1(\mu, E)$. The fact that

$\|\int f\,d\mu\| \leqq \int \|f(x)\|\,d\mu x$ can be rephrased: the Bochner integral B_μ is a linear function on $L_1(\mu, E)$ of norm at most one (of course $\|B_\mu\| = 1$; consider a function $x \mapsto \chi_A(x)a$). We show that $L_1(\mu, E)$ is complete. The critical fact: the inequality $\|\int f\,d\mu\| \leqq \int \|f(x)\|\,d\mu x$ permits us to deduce convergence in E from the theorems about convergence for scalar valued functions.

23 THEOREM *If E is a Banach space and μ is a measure, then each swiftly convergent sequence in $L_1(\mu, E)$ converges pointwise a.e. and in norm to a member of $L_1(\mu, E)$, and consequently $L_1(\mu, E)$ is complete.*

PROOF It is sufficient to show that if $\{f_n\}_n$ is a sequence in $L_1(\mu, E)$ such that $\sum_n \|f_n\|_1 < \infty$, then there is a member f of $L_1(\mu, E)$ such that $lim_n \|f(x) - \sum_{k=0}^n f_k(x)\| = 0$ for μ a.e. x and $lim_n \|f - \sum_{k=0}^n f_k\|_1 = 0$. But $\sum_n \|f_n\|_1 = lim_n \sum_{k=0}^n \int \|f_n(x)\|\,d\mu x = lim_n \int \sum_{k=0}^n \|f_n(x)\|\,d\mu x$ and so, by B. Levi's theorem, the increasing sequence $\{x \mapsto \sum_{k=0}^n \|f_k(x)\|\}_n$ converges μ almost everywhere to an integrable function g. Let $f(x) = \sum_n f_n(x)$ for points x such that $\sum_n \|f_n(x)\| = g(x)$ and let f be 0 otherwise. Then f is easily seen to be σ-simple because each f_n is and $f \in L_1(\mu, E)$ because $\int \|f(x)\|\,d\mu x \leqq \int g(x)\,d\mu x$. Moreover $lim_n \|f - \sum_{k=0}^n f_k\|_1 = lim_n \int \|f(x) - \sum_{k=0}^n f_k(x)\|\,d\mu x = 0$ by the dominated convergence theorem. ∎

Finally, a generalization of the dominated convergence theorem is valid for E valued functions.

24 DOMINATED CONVERGENCE THEOREM *Let $\{f_n\}_n$ be a sequence in $L_1(\mu, E)$ such that $\{f_n(x)\}_n$ is a Cauchy sequence for μ a.e. x and suppose g is an integrable nonnegative function such that $\|f_n(x)\| \leqq g(x)$ for each n and μ a.e. x.*

Then there is a member f of $L_1(\mu, E)$ such that $f = \lim_n f_n$ a.e., $lim_n \|f_n - f\|_1 = 0$, and consequently $\lim_n \int f_n\,d\mu = \int f\,d\mu$.

PROOF The function f defined by $f(x) = lim_n f_n(x)$ if the limit on the right exists and zero otherwise, is σ-simple in view of proposition 20 and $\|f(x)\| \leqq g(x)$ for μ a.e. x. Consequently the function $x \mapsto \|f_n(x) - f(x)\|$ is an $\mathscr{L}\mathscr{A}$ measurable function with a support in $\mathscr{A}_\sigma$ which is dominated μ a.e. by $2g$. By the dominated convergence theorem for scalar functions, $lim_n \|f_n - f\|_1 = lim_n \int \|f_n(x) - f(x)\|\,d\mu x = \int lim_n \|f_n(x) - f(x)\|\,d\mu x = 0$. ∎

SELECTED REFERENCES

C. D. ALIPRANTIS and O. BURKINSHAW
[1] *Principles of Real Analysis*, North-Holland (1981)
E. ASPLUND and L. BUNGART
[1] *A First Course in Integration*, Holt, Reinhart and Winston (1966)
T. A. BOTTS and E. J. MCSHANE
[1] *Real Analysis*, Van Nostrand (1959)
D. L. COHN
[1] *Measure Theory*, Birkhäuser (1980)
G. B. FOLLAND
[1] *Real Analysis*, John Wiley and Sons (1984)
P. R. HALMOS
[1] *Measure Theory*, Van Nostrand (1950)
E. HEWITT and K. STROMBERG
[1] *Real and Abstract Analysis*, Springer-Verlag (1965)
K. HOFFMAN
[1] *Analysis in Euclidean Spaces*, Prentice-Hall Inc. (1975)
A. MUKHERJEA and K. POTHOVEN
[1] *Real and Functional Analysis*, Parts A and B, Plenum Press (1986)
K. R. PARTHASARATHY
[1] *Introduction to Probability and Measure*, Springer-Verlag (1978)
W. F. PFEFFER
[1] *Integrals and Measures*, Marcel Dekker, Inc. (1977)
A. C. M. VAN ROOIJ and W. H. SCHIKHOF
[1] *A Second Course on Real Functions*, Cambridge University Press (1982)

H. L. ROYDEN
[1] *Real Analysis*, The Macmillan Co. (1968)
W. RUDIN
[1] *Real and Complex Analysis*, McGraw-Hill (1987)
S. SAKS
[1] *Theory of the Integral*, Stechert (1937)
I. SEGAL and R. A. KUNZE
[1] *Integrals and Operators*, McGraw-Hill (1968)
K. T. SMITH
[1] *Primer of Modern Analysis*, Springer-Verlag (1983)
K. STROMBERG
[1] *An Introduction to Classical Real Analysis*, Wadsworth International (1984)
A. J. WEIR
[1] *Lebesgue Integration and Measure*, Cambridge University Press (1973)
R. L. WHEEDEN and A. ZYGMUND
[1] *Measure and Integral*, Marcel Dekker, Inc. (1977)
A. C. ZAANEN
[1] *Integration*, North Holland (1967)

ADDITIONAL TITLES

W. ARVESON
[1] *An Invitation to C*-Algebras*, Springer-Verlag (1976)
K. BICHTELER
[1] *Upper Gauges and Integration of Linear Maps*, Springer-Verlag, Lecture Notes (1973)
A. BROWN and C. PEARCY
[1] *Introduction to Operator Theory I*, Springer-Verlag (1977)
C. CONSTANTINESCU, K. WEBER and A. SONTAG
[1] *Measure and Integral*, Vol. I, John Wiley and Sons (1985)
R. ENGELKING
[1] *General Topology*, PWN, Warszawa (1977)
D. H. FREMLIN
[1] *Topological Riesz Spaces and Measure Theory*, Cambridge University Press (1974)
K. JACOBS
[1] *Measure and Integral*, Academic Press (1978)
J. L. KELLEY
[1] *General Topology*, Springer-Verlag (1975)
J. L. KELLEY and I. NAMIOKA *et al.*
[1] *Linear Topological Spaces*, Springer-Verlag (1976)
C. KURATOWSKI
[1] *Topologie* I, Warszawa (1933)

J. Mikusinski
[1] *The Bochner Integral*, Birkhäuser (1978)
L. Nachbin
[1] *The Haar Integral*, Van Nostrand (1965)
J. C. Oxtoby
[1] *Measure and Category*, Springer-Verlag (1971)
F. Riesz and B. Sz-Nagy
[1] *Functional Analysis* (English ed.), Frederick Ungar (1955)
M. H. Stone
[1] *Notes on Integration* I–IV, Proc. Nat. Acad. Sci. U.S.A. (1948, 1949)
F. Topsöe
[1] *Topology and Measure*, Springer-Verlag Lecture Notes (1970)
A. Weil
[1] *L'integration dans les groupes topologiques et ses applications*, Actualités Sci. Ind. (1940)

INDEX

9 780387 966335